CHENIA

隐花植物生物学

CONTRIBUTIONS TO CRYPTOGAMIC BIOLOGY

第三-四卷 Vol. 3-4

主 编: 吴鹏程 贝尼托.陈 贾 渝

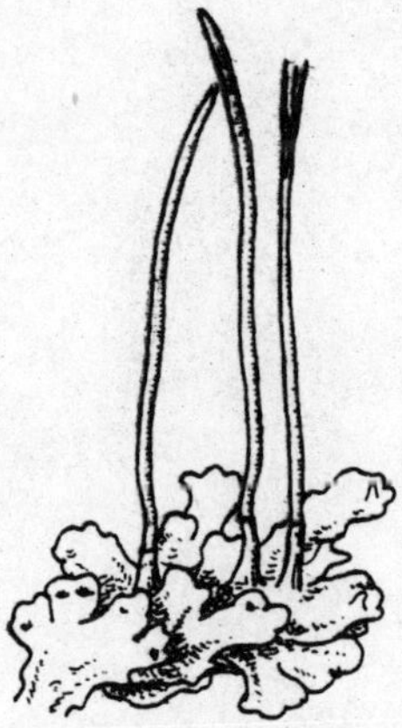

知識出版社

·北京·

责任编辑： 徐 扬

封面设计： 郭木森

图书在版编目（CIP）数据

CHENIA隐花植物生物学 第三-四卷/吴鹏程等主编. -北京：知识出版社，1997.3

ISBN 7-5015-1338-4

Ⅰ.C… Ⅱ.吴… Ⅲ.孢子植物-生物学-研究报告 Ⅳ.Q949.1

中国版本图书馆CIP数据核字(97)第02214号

知识出版社出版发行

(北京阜城门北大街17号 100037)

中国科学院植物研究所印刷

开本：787×1092 1/16 印张：11

1997年3月第一版 1997年3月第一次印刷

印数：1-500 字数：243000

CHENIA

CONTRIBUTIONS TO THE CRYPTOGAMIC BIOLOGY Vol. 3–4

目 录

研究报告

CHENIA

CONTRIBUTIONS TO THE CRYPTOGAMIC BIOLOGY Vol. 3–4

CHENIA

CONTRIBUTIONS TO THE CRYPTOGAMIC BIOLOGY Vol. 3–4

Contents

Research reports

CHENIA

CONTRIBUTIONS TO THE CRYPTOGAMIC BIOLOGY Vol. 3–4

中国—北美苔藓植物地理分布关系

贾 渝 吴鹏程
（中国科学院植物研究所，北京，100093）

PRELIMINARY STUDY ON THE RELATIONSHIP OF BRYOFLORAS BETWEEN CHINA AND NORTH AMERICA

Jia Yu and Wu Pan-cheng
(*Institute of Botany, Academia Sinica, Beijing 100093, China*)

Abstract: Since the middle of ninteen century, Asa Gray and many others have discussed the relationship of the vascular floras between East Asia and north America. Later, bryologists such as Sharp, Iwatsuki and Schofield compared the moss floras between Japan and North America. In 1980, *Takakia*, a genus known from Tibet, SW China, Japan and western North America, was considered to be the most primitive group of bryophyte. Its present distribution indicates a close relationship of the bryofloras between China and North America. Wu and Magill in 1990 compared the bryofloras of the *Metasequoia* region in western Hubei, central China, with that of the *Taxodium* forest in south eastern United States, and suggested that the migration of whole floras between the two continents was possible.After the Tertiary, the diversification of the species of bryophytes between the two continents became apparent due to the separation of Eurasia and North America continents and the influence of glaciation, although temperate elements still predominate in both continents. Considering the present day bryofloras of these two continents, the similarity coefficiency of moss species between central China and western North America and eastern North America are about 26.6% and 32.4% respectively. The similarity coefficiency exhibited at the family and generic levels is even higher. It may be assumed from the above relationship that the floras of the two continents probably diverged 50 million years ago.

Key words: China; North America; Similarity coefficiency

摘要：自从19世纪中期Asa Gray论述了中国和北美被子植物之间的关系之后，中外学者相继对两地区间的被子植物的一些类群作了研究，发现了它们的相似性，并认为两大陆之间的植物区系成分曾发生相互迁移．本世纪50年代之后，日本、美国等国学者就东亚和北美苔藓植物区系的关系作过探讨．本文就中国和北美苔藓植物之间的区系关系作了简要的讨论，认为两地区的苔藓植物区系有着共同的来源．由于冰川及气候变化的影响，加之两地间自然地理条件的差异，中国和北美苔藓植物区系产生了一些差异．

关键词：中国；北美；苔藓植物地理分布

19世纪中期，东亚和北美被子植物之间的关系由Asa Gray作了论述．在此之后，胡先骕(1935）、李惠林（1952）等研究中国和北美两地区间木本植物等类群，发现了它们的科、属和种间的相似性，认为两大洲种子植物之间曾有相互迁移发生．然而迄今为止，在苔藓植物方面仅Sharp (1953, 1972)、Iwatsuki (1958) 和Schofield（1965）就北美和日本间苔藓植物区系的关系作过一些探讨，以此为依据进而简要地讨论了东亚和北美之间的苔藓植物区系关系．当时可能因对于中国苔藓植物区系的资料了解甚少而未提及中国,而中国苔藓植物学在这方面的专门性研究基本上属于空白．1983年，被认为最原始的苔藓植物目—藻苔目Takakiales在中国西藏的发现，及随后在中国横断山区又发现了该种植物(汪楣芝, 1993)的分布，揭示了中国和北美之间重要苔藓植物类群的密切关系(吴鹏程等,1983)．1990年, 吴鹏程和Magill首次用中国湖北水杉林*Metasequoia*和北美东南部池杉*Taxodium*林内苔藓植物类群间的相似性加以论述，证实了东亚与北美间的渊源．中国是东亚自然地理范围内重要组成部分，有必要对它与北美之间的苔藓植物区系关系作一些初步的讨论．

一、中国和北美之间的苔藓植物区系的比较

正如上所述，中国和北美种子植物间的区系存在较为密切的关系．我们在研究两地间苔藓植物区系资料时，发现科和属级水平的相似性很高，但存在一定的差别．以藓类植物为例，分布于北美的光藓科Schistostegaceae,目前在中国尚无记录，中国有记录的扭叶藓科Trachypodaceae、金毛藓科Myuriaceae、四齿藓科Georgiaceae、树生藓科Erpodiaceae、毛藓科Prionodontaceae、带藓科Phyllogoniaceae、白藓科Leucomiaceae、刺果藓科Symphyodontaceae、树灰藓科Hypnodendraceae等科在北美地区尚无分布．这些北美地区没有分布的科大都限于热带、亚热带范围．

两地区之间属的相似性较高，其相似性系数约为54％．中国藓类植物共有约409属（Redfearn et Wu,1986）其中有106属是北美地区没有记录的．这106属中大致包括以下几种类型：1）中国或东亚地区特有属，如拟牛毛藓属*Ditrichopsis*、拟短月藓属*Brachymeniopsis*、兜叶藓属*Horikawaea*等为中国特有，而新船叶藓属*Neodolichomitra*、褶藓属*Okamuraea*等则为东亚特有属．2）分布于热带及亚热带地区范围的属，台湾藓属 *Taiwanobryum* 和蕨藓属 *Pterobryon* 是极好的例证．种类较少的属，山毛藓属 *Oreas*、石芽藓属 *Stegonia* 及侧立藓属*Pleuroweisia*等按其地理分布范围应属于温带成分．而处于温带气候下的北美地区却未见分布, 其原因将在本文的后部作简要地讨论．

北美的苔藓植物中有许多属只有1个种或极少数种在本地区分布，以北美东部为例，仅有1种的属就有103属．在北美东部地区属和种数之比为1: 3.2(Lane,1985)．

以中国中部的神农架为例，两地之间种的相似性相对较低，分别是：与北美东部的相似性为32.4%，与北美西部的相似性为26.6% (Wu, Crosby & Magill, 1983). 经比较发现，北美与中国共有的苔藓植物种类仍以温带分布为主. 泥炭藓属*Sphagnum*有42种，其中有21种是两地共同分布的；紫萼藓属*Grimmia*有24种及2变种，14种和

表1. 中国和北美地区热带、亚热带分布藓类科的比较

Tab.1. The calculation of the tropical and subtropical families of mosses between China and North America

Region	China			North America		
Number of Taxon	Genus	Species	Infra-species	Genus	Species	Infra-species
Leucobryaceae	4	15	1 f	2	3	
Fissidentaceae	1	45	2 subsp, 7 var, 1 f	1	27	1 f
Calymperaceae	4	27	1 subsp	2	10	
Prionodontaceae	1	1				
Trachypodaceae	5	11	10 var, 2 f			
Pterobryaceae	10	29	4 var, 2 f			
Meteoriaceae	11	42	9 subsp, 14 var	3	3	
Neckeriaceae	10	58	10 var	5	7	
Hookeriaceae	10	29	2 var	6	6	
Hypopterygiaceae	4	20		1	1	
Sematophyllaceae	22	78	3 subsp, 1 var	6	10	

subsp. =subspecies; var.=variety; f.=form

2变种是共有的种类；真藓属*Bryum*有25种和1变种，其中18种为两地共有.

在中国和北美共有种中热带成分极少，花叶藓属*Calymperes*在北美仅有4种，共有分布的仅1种，而本属全世界有100种，网藓属*Syrrhopod*on在北美分布6种，与中国共有的3种；油藓科Hookeriaceae共约500种，在北美仅存在6种，而与中国共有分布的仅尖叶油藓*Hookeria acutifolia*；锦藓科 Sematophyllaceae 约有700种，在北美分布的种类仅10种，两地共有的6种. 北美大陆的热带苔藓植物其数量是非常稀少的，主要分布在临近中南美洲的狭小范围之内，它们主要是与中南美洲地区的热带苔藓种类存在较为密切的关系. 中国的热带、亚热带苔藓植物的种类明显多于北美(见表1). 这些种类的地理分布多与东南亚地区存在密切关系. 两地热带分布的苔藓植物无论是数量和来源上都有差别.

在中国和北美地区共有的种类中，除了世界广泛分布以及北半球广布或北温带分布等，尚有不少种类是间断性地分布于中国和北美两地，这种分布类型无疑为中国和北美之间区系渊源关系提供了极好的例证．例如：异枝皱蒴藓*Aulacomnium heterostichum*，分布于中国湖北、川西高山地带、北部长白山地区和日本，同时也见于北美东部至加拿大；残齿藓*Forsstroemia trichomitria*广泛地分布于北美，中国吉林和福建也有分布；皱叶牛舌藓*Anomodon rugelii*见于中国秦岭和东部沿海诸省，北美东部、北部，南至佐治亚州和田纳西州；刺叶小鼠尾藓*Myurella sibirica*分布于中国北部及西南高山地带，从北美的东北部到阿拉斯加，南至北美的东部及东南部也有分布；多疣悬藓*Barbella pendula*从中国东南部至北美、中美洲；粗枝藓*Gollania turgens*见于阿拉斯加、加拿大西部、俄罗斯远东地区、中国大陆及日本：*Orthotrichum sordidum*在中国东北部、朝鲜、日本和北美，通常是附生状态，偶尔生长于岩面，海拔分布从海平面直至1700m；最近，在中国新疆天山东部发现的钝尖藓 *Calliergidium bakeri* (Ren.) Grout（罗健馨、赵建成，1991）原仅 见于北美大陆．

表2. 中国和北美分布的“姊妹种”

Tab.2 The sibling species of mosses between China and North America

China	North America
Homaliadelphus sharpii var. *rotundatus*	*Homaliadelphus sharpii* var. *sharpii*
Climacium japonicum	*Climacium america*
Venturiella sinensis var. *sinensis*	*Venturiella sinensis* var. *angusti-annulata*
Drummondia sinensis	*Drummondia prorepens*
Entodon sullivantii var. *sullivantii*	*Entodon sullivantii* var. *versicolor*
Diphyscium involutum	*Diphyscium cumberlandianum*
Ascidiota blepharophylla subsp. *blepharophylla*	*Ascidiota blepharophylla* subsp. *alaskana*

此外，仅分布于两大陆块之间的“姊妹种”（表2）更进一步地阐明了东亚和北美之间的历史渊源．

二、影响中国和北美苔藓植物地理分布的因素

1、植物区系的差异

在古生代，古地中海曾将今日的亚洲大陆分隔成南北两部，形成亚欧大陆两个基本植物区的分界，北面为泛北极植物区，南侧为古热带植物区（刘德生等，1986）．古热带植物区形成以后，尽管经历了漫长的地质时期，但变化不大．不少地区仍保存起源于中生代的“活化石”，例如树蕨、银杏、苏铁等．泛北极植物区在发展过程中由于海水进退、高山隆起、沙漠形成以及第四纪大冰川等较大的地质变迁，使该区的植物区系发生了极大变化，因此，今日泛北极区的植物界是较为年轻的，植物种类亦不如古热带区复杂．

东亚是地球上热带植物区系可以过渡为温带植物区系的唯一地区．因为这里南部没有高大山脉，也没有荒漠（荒漠是热带和温带喜湿植物不能通过的），同时又有很大的湿度，因此在这一地区热带植物区系逐渐过渡为温带植物区系，中国作为东亚的主要组成部分，植物种类(包括苔藓植物)非常丰富，而且植物区系成分具有古老性．与此同时，大洋洲板块在第三纪迅速向亚洲板块接近以及印度板块向北“漂移”与亚洲相撞，都给予冈瓦纳古陆植物向劳亚古陆渗透增添了机会．前者通过马来西亚地区对中国华南地区产生影响，后者对中国西南地区和华南地区产生影响(吴鹏程、林邦娟，1978)，因此，中国的苔藓植物区系成分中不仅含有泛北极成分，也有与热带亚洲密切相关的热带亚洲成分，同时，还存在一些来自南半球的区系成分．中国由于广阔的地域、复杂的地理和气候条件更孕育了中国丰富的物种多样性．即使同一种在中国不同的地区也可能发生一些形态上的差别，如：多瓣苔属*Macvicaria*在横断山区的种类与中国东北地区的种类在形态上出现了差异（罗健馨、汪楣芝，1983）．

北美大陆的苔藓植物区系是以温带分布类型的泛北极成分为主．这些植物主要分布在北美大陆的北部和中部．北美南部联接南美大陆的地域极为狭窄，南马雷德山横亘其北部，对热带植物向北传播起着阻碍作用，使这里成为泛北极植物区系和新热带植物区系的明显分界线．

2、气候变化的影响

北美大陆经历了四次大冰期及三次间冰期的演变，它的山体又多为南北走向，冰川到来时，大量的物种绝灭或向南退却．待冰川退却时，一些“幸存”的种类向北推进，但这些种类由于适应新的环境而发生了一些变化，有的形成了新的物种．这也许是为什么许多属在北美地区仅有1种或极少种类的原因之一．在前面我们曾经提到有些分布于温带气候下的属在北美却没有分布．作者认为这可能是由于受冰川的影响而导致整个属的绝灭，但一些包含较多种类的属在北美大陆则可保留少数的种．包括中国在内的东亚山体走向复杂，受冰川的影响较小．例如分布于中国和北美两地的山区的白氏藓*Brothera leana*几乎不产生孢子体，却有很多的芽胞．该种可能是由于冰川的影响使其逐渐过渡到完全的无性生殖来维持种群的生存，但最近从四川西部采集到该种标本，通过观察发现其芽胞的形态已与北美的种类有所差别．据研究推测，这一地区在第四纪冰期是植物避难所之一（齐金根，1991）．一般而言，大多数苔藓植物

的进化是极其缓慢的，然而如遇大的地理和气候的变迁（如：造山运动、海浸和冰期等）也会加快物种的分化，使其两地区的区系成分产生差异．

东亚和北美阿巴拉契亚山地区代表劳亚古陆的两个主要范围，在此范围之内植物“逃避”了对植物影响深远的冰川期，而且它们又有足够的湿度来维持大量在此“避难”的物种．相比之下，欧洲就失去一些在上述两地保存下来的分类群（如日鳞苔属*Nipponolejeunea*等）．一般来说，在东亚较为温和的渐新世气候下，许多属和种被保存下来；在北美大陆则只有一定数量的种被保存下来；而在欧洲大陆大量的物种遭到绝灭．因此，许多东亚—北美分布的种类在很大程度上是由于该种在其他地区绝灭而造成的结果．

在气候发生较大变化时，苔藓植物的生殖方式会有变化，一般是由有性生殖向无性生殖方式转变，并以专门的繁殖体来进行．从藓目、变齿目、灰藓目、葫芦藓目和曲尾藓目有较多的种有专门的繁殖体．例如，在南极地区的藓类植物主要以无性生殖的繁殖方式进行．同时，显而易见的是，在苔藓植物中雌雄异株的种类受冰川的影响比雌雄同株要大．雌雄异株的种类中能适应新环境并产生适当变异的种类，保留下来并可能形成新分类群，不能适应的种类虽然能保留下来，但被限在狭窄的范围之内，有些种类显然是被灭绝的．钟帽藓*Venturiella sinensis* (Rabenh.) C. Muell.是值得注意的，它是雌雄异株的种类，分布区范围仅限于中国、日本等东亚地区以及北美地区．在北美大陆其种群小，而且植物体也细小，它的孢蒴上的环带较为狭窄（57 - 69 μm），环带细胞呈等径状；而东亚的种类则是69–129μm，且环带细胞呈长形或椭圆形，使其足以从原种中分离出来，产生1个新的变种：*Venturiella sinensis* (Rabenh.) C. Muell. var. *angustiannulata* Griff. & Sharp．这是否由于冰期气候的巨变而引起的种群分化，现在还不能完全肯定（Iwatsuki, 1992）．

在苔藓植物中,顶蒴藓类比侧蒴藓类的染色体的数目具更多的变异．从藓科Pottiaceae、真藓科Bryaceae、葫芦藓科Funariaceae倾向于生活在那些裸露而易受到干扰的生境，在这样的生境之中，孢子体外部损伤可以诱导出同源多倍体．而侧蒴藓类倾向于占领较稳定而温和的环境，相对而言，在此孢子体外部受伤的机会很少．因此，我们推断在冰期的影响之下，北美地区分布的侧蒴藓类受到的影响要大于顶蒴藓类．

三、起源问题

通过以上一些事实，表明中国与北美的苔藓植物有着共同的起源．目前苔藓植物学家都接受这样的观点：即长距离的孢子传播在苔藓植物的迁移所起的作用是极为有限的．苔藓植物可能会很好地适应微小的生境，对大气候的变化的抗逆能力比维管植物强，能够呈现一些孑遗分布，当其他高等植物面临绝灭时能进一步演化．在北美东部的阿巴拉契亚山存在大量古老的起源于北极地区的区系成分的中生植物．在北美特有属极少，且几乎没有一个是狭窄分布,而明显地呈孑遗分布状态．缺乏特有属说明北美 苔藓植物区系的起源是古老的．例如，拟扁枝藓*Homaliadelphus sharpii* (Williams) Sharp明显地来源于热带的成分，如今限制分布于东亚和北美地区．湿鳞苔属*Hygrole-*

*jeunea*在阿拉斯加被发现，呈现残遗分布，显然是由于生活在未受冰川覆盖地区而保存下来的，它可能是温带或热带的区系成分(Sharp,1972)。然而,以中国和日本为主体的亚洲东部地区由于地史及气候条件等影响，迄今仍为保留有世界上特有属数（中国约有35个，日本14个）最多的地区之一（Deguchi et Iwatsuki, 1984; Wu. 1992）.

实际上, 在第三纪之前，苔藓植物的大类群就已经开始分化. 许多科也明显地分化，但种数却很少，甚至有些是单属科. 广布种显然是在以前就已经广泛分布并且几乎一直未变. 在苔藓植物中特有现象无疑是古老的，因为苔藓植物的构造相对简单，要使其在一个特殊生境之中突然爆发式地产生大量的新种是不可能的.

参考文献

[1] 齐金根、陈伟烈、王金亭、阎建平，1991: “华西野生植物实验保护中心”及其邻近地区的植物区系与植被特征分析. 植物学集刊: 87-105.

[2] 刘德生等，1986: 《世界自然地理》101-102页，北京，高等教育出版社.

[3] 汪楣芝，1993: 横断山区苔藓植物新分布 CHENIA 1:75-82.

[4] 吴鹏程、林邦娟，1978: 海南苔类植物区系与地理分布的初步讨论. 植物分类学报16(2):56-71.

[5] 吴鹏程、罗健馨、汪楣芝，1983: 一个原始苔类的目—藻苔目在中国的发现. 植物分类学报 21(1):105-107.

[6] 罗健馨、汪楣芝，1986: 横断山脉苔藓植物特有属和新记录初报. 横断山研究专集 442-452页. 北京，科学技术出版社.

[7] 罗健馨、赵建成，1991: 中国苔藓植物的一个新记录属—钝尖藓属. 新疆大学学报（自然科学版）: 70-71.

[8] Deguchi, H.et Z.Iwatsuki, 1984: Bryogeographical relationships in the moss flora of Japan. *Hourn. Hattori Bot. Lab*. 55:1-11.

[9] Iwatsuki,Z.,1992: The moss flora of Japan and its North American connections *Bryobrothera* 1:1-7.

[10]Lane, D.M. 1985: A Quantitative Study of the Mosses of Eastern North America. Monogr.Syst. Bot. Missouri Bot. Gard. 11:45-50.

[11]Schofield, W. B 1965: Correlations between the moss floras of Japan and British Columbia. Canada. *Journ. Hattori Bot*. Lab. 28:17-42.

[12]Sharp,A.J.,1972:Phytogeographical Correlations between the Bryophytes of Eastern Asia and North America. *Journ. Hattori Bot. Lab*. 35:265-267.

[13]Wu P.C.et R.E.Magill,1990: A Comparison of Bryofloras of the *Metasequoia* Region,Western Hubei, China and the *Taxodium* Forests of the Southeastern United States. *Acta Bryolichenologica Asiatica* 2(1,2):33-40.

[14]Wu P.C.,1992: The East Asiatic genera and endemic genera of the bryophytes of China. *Bryobrothera* 1:99-117.

坦桑尼亚西南部的*GANGAMOPTERIS*和*GLOSSOPTERIS*植物化石 *

胡雨帆 徐仁
(中国科学院植物研究所，北京 100093)
何德长
(中国煤炭科学院地质勘探所，西安 710054)

FOSSILS OF *GANGAMOPTERIS* AND *GLOSSOPTERIS* IN SOUTH－WESTERN TANZANIA

Hu Yu-fan and Hsu Ren
(*Institute of Botany, Academia Sinica, Beijing 100093, China*)
He De-chang
(*Institute of Geology, Chinese Academy of Coal Xi'an 710054, China*)

Abstract: The Coalfield Investigation Team of the People's Republic of China undertook an investigation of some coalfields in the south－western part of the United Republic of Tanzania.A report on the main results of the investigation has been submitted to the Tanzanian Government.

In the past, some western geologists carried out investigation and study on the geological formations of the coalficlds in the southwestern part of Tanzania.

The term "Karroo System" or "Songea Series" employed by Stockley in 1930 has been generally used for more than 30 years. Although further research work has been done by others in recent years, no new insight beyond the old frame-work has been achieved.

The fact that the Investigation Team has found an abundance of plant fossils in the "Siltstone Stage" in the Ketewaka-Mchuchuma Coalficld negates the view held for the past few decades that there was no plant fossil or that it was extremely hard to find a fossil in this formation.

The Investigation Team examined and analysed more than 100 specimens of plant fossils (in which more than 30 specimens were well-preserved and could be studied by authors).

Through a general study,the Investigation Team is interested in the fact that the plant genus and species found in the Siltstone Stage are represented mainly by *Gangamopteris* and that in spite of the presence of some elements of Early Permian age such as *Palaeovittria tanzaniensis* the genus *Glossopteris* was not found in this Stage.

---

* 本文系国家自然科学重点基金(No.39230030)和面上基金(No.39470048)资助项目，并得到中国科学院古生物学科特别支持费和煤炭部科研特别支持费资助。

There are nine species and two varieties of fossil plants occrring in the Siltstone stage of the Idusi Series, namely: *Gangamopteris cyclopteroides* Feistm., *G.* cf. *cyclopteroides* Feistm., *G. cyclopteroides* cf. var. *attenuata* Feistm., *G. cyclopteroides* var. *acuminata* Feistm., *G.* cf. *fibrosa* Maithy,*G.sp.*(sp.nov.?), *G.sp.A,G.sp. B*, *Palaeovittaria tanzaniensis* (sp.nov.)., *Noeggerathiopsis hislopi* (Bunbury) Feistm. and the equisctaceous stem. Among these, *Gangamopteris* is predominant in quantity and the equisetaceous stem and *Noeggerathiopsis* are in the second place. *Palaeovittaria tanzaniensis* is extremely rare, and only one piece of fossil has so far been found.

But *Glossopteris indica* and *Glossopteris browniana* have been found in the Mhukuru Series of Early Permian age.

Key Words: Tanzania; Plant fossils; *Glossopteris; Gangamopteris.*

摘要: 本文报道中国煤矿地质考察组对坦桑尼亚联合共和国西南部一些煤田进行考察时，采集的植物化石共计12种2变种，这是一类特殊的南大陆冈瓦纳植物群，其主要是*Gangamopteris*类型，还有2种*Glossopteris*. 地质时代属于石炭二叠纪，为在该煤地层中首次发现，这一植物群与中国华夏植物群截然不同，其不仅有重要的地层意义，同时也有着重要的古植物地理意义.

关键词: 坦桑尼亚; 植物化石; *Gangamopteris; Glossopteris*

中国煤矿地质考察组曾对坦桑尼亚联合共和国西南部一些煤田进行过考察. 其主要考察成果向坦桑尼亚政府提交了报告.

在此之前，曾有一些西方地质学者对坦桑尼亚西南部各煤田地层进行过调查研究. 但没有找到可供鉴定的化石，而一直沿用斯托克莱1930年的命名，笼统地把该地区的地层称之为卡鲁系或桑格亚系. 近年来，虽也有人进一步作了一些工作，但未见任何化石之报道.

中国煤矿地质考察组发现粉砂岩段盛产植物化石是在基德华加一姆丘丘玛煤田，这一发现打破了过去几十年来一直认为这段地层中“不产”或“极难找到”植物化石的说法.

考察组在坦桑尼亚西南部所采集的植物化石标本共100余块(其中保存较好可供研究者有30余块）由煤炭部交我们研究鉴定，特别对其中30余块化石作了认真研究和详细描述，并作出了文字报告. 近年来，研究南大陆二叠纪及板块的学者纷纷要求提供这批资料，为加强国际交流，并按徐仁先生临终之嘱，现将其系统整理后公开报道.

研究中，我们在依多西组粉砂岩段见到以众多的恒河羊齿*Gangamopteris**为代表的植物群，尽管也有一些二迭纪的分子如坦桑古书带叶（新种）*Palaeovittaria tanzaniensis* sp. nov. 出现, 但未见到舌羊齿（详细名单见后）. 在基德华加—姆丘丘玛煤田的尼亚克莱河见到印度舌羊齿*Glossopteris indica*，在姆呼库鲁煤田的姆呼库鲁河找到了均网舌羊齿*Glossopteris browniana*和脊椎木种*Vertebraria* sp. 等，它们都属于上二迭统的姆呼库鲁组的地层. 这是非常有意义的发现. 本文对以往的地层划分提出了建议，相信这将有助于地层问题的讨论.

植物化石出现于依多西粉砂岩段的共有9种和2变种:

木贼类茎 Equisetaceaeus stem.

圆恒河羊齿 *Gangamopteris cyclopteroides* Feistm.

似楔形恒河羊齿 *G. cyclopteroides* cf. var. *attenuata* Feistm.

似圆恒河羊齿 *G.* cf. *cyclopteroides* Feistm.

渐尖恒河羊齿 *G. cyclopteroides* var. *acuminata* Feistm.

似条纹恒河羊齿 *G.* cf. *fibrosa* Maithy

恒河羊齿1种(?新种)G. sp.(? sp. nov.)

恒河羊齿A种G. sp.A

恒河羊齿B种G. sp.B

坦桑古书带叶(新种)*Palaeovittaria tanzaniensis* Hu et He (sp. nov.)

辐脉苛达(新组合)*Cordaites hislopi*(Feistm.) Hu, comb. nov.

在这些化石中，以恒河羊齿为最多，其次为木贼类茎和辐脉苛达两种化石．古书带叶是一种极为少见的化石，而且仅仅见到一块．

本文主要内容包括以下几个方面:

一、较详细地介绍了粉砂岩段中所发现的以恒河羊齿为代表的冈瓦纳植物属种的特征，介绍了一个植物化石新种—坦桑古书带叶*Palaeovittaria tanzaniensis* Hu et He (sp. nov.) 和一个新组合—辐脉苛达*Cordaites hislopi* (Feistm.) Hu comb. nov.

二、从沉积角度来看，依多西组底部砂岩段属冰川泥砾岩；泥岩段具典型的纹层，被视为冰湖沉积物；粉砂岩段则以其含有较多的植物化石为特征，岩性比较稳定，是一般湖沼相沉积物．这说明当时的气候逐渐变得温暖起来，但是与形成上部含煤建造的气候相比较，仍然比较寒冷．因此，粉砂岩段具有过渡层的特性，甚至可以被视为石炭纪—二迭纪的过渡层．

三、若从植物化石来看，粉砂岩段中所产植物虽有二迭纪的面貌，但多为石炭纪的成分．

四、下姆丘丘玛组含具有经济价值的煤层，依多西组则不含煤．从坦桑尼亚本国的地层系统和生产观点来考察，将粉砂岩段这样一个过渡层依多西组上部，暂时列为上石炭统是较为适宜的．本文还详细介绍了下二迭统下姆丘丘玛组中的舌羊齿未定种和上二迭统姆呼库鲁组中的印度舌羊齿*Glossopteris indica* Schimper和均网舌羊齿*Glossopteris browniana* Brongn等重要冈瓦纳植物分子．

DESCRIPTION OF FOSSILS

Equisetaceaous stem.

Pl. I, fig.1.

The equisetaceaous stem is the largest one of three fragmentary specimens in present material. It is 7.8cm. long and 2.2cm. broad, and bearing clear nodes and internodes. On the stalk,ribs and furrows are alternately arranged and continuing through the nodes and with fine longitudinal striae. The number of ribs is about 7 in a centimetre.

Locality: The Mchuchuma River, Ketewaka-Mchuchuma Coalfield.

Horizon: The Siltstone of the Idusi Series.

Gangamopteris cyclopteroides Feistm.

Pl. I, fig. 2.

It is a fragmentary piece of a leaf impression, which measures 11.3cm. long and 5cm. broad. Owing to the incomplete nature of the specimen, no idea can be formed about its shape, apex and base. The midrib is absent.The veins radiate out from the base, bifurcate, slowly bend outwards, and anastomose with each other into network.Towards the central portion of the leaf the meshes are broader.

According to the partial leaf shape and venation, this specimen demonstrates the characters of *Gangamopteris cyclopteroides* which is very common in the Upper Carboniferous and Lower Permian. There is no much doubt that it would be referred to *Gangamopteris cyclopteroides*.

Locality: The Mbamba-Bay Coalfield.

Horizon: The Siltstone of the Idusi Series.

Gangamopteris cf. *cyclopteroides* Feistm.

Pl. I, figs. 4-5

The two fragmentary specimens, both representing leaf bases,probably belong to *Gangamopteris cyclopteroides*. The one shown in fig.4 is larger,7.5cm. long and 2.2cm. broad. The midrib is absent. The veins radiate out from the base, dichotomous, anastomose to form oblong meshes which are broader and longer towards the centre than the margin.

Locality: The Mbamba-Bay Coalfield.

Horizon: The Siltstone of the Idusi Series.

Gangamopteris cyclopteroides var.*acuminata* Feistm.

Pl. II , fig.2.

In present material, only one poorly preserved specimen represents this variety. The leaf if rhomboidal, 9cm. long and 2.5 cm. broad, with a bluntly pointed apex. The leaf base is very narrow as compared with its middle portion. The median region is occupied by a group of subparallel veins, from these give off the slightly arched secondary veins which are dichotomous, anastomose with each other into narrow and long meshes.

The leaf is characterized by the rhomboidal shape and the very narrow base. The present specimen agrees in all respects with *Gangamopteris cyclopt* eroides var. *acuminata* (Feistmantel, O.1886,p.32, Pl.7A,fig.4; Pl.8A, fig.5; Pl.11A, figs.4-7) from the

Talchir Group of Karanpura and Auranga Coalfields of India. There is no doubt that it would belong to the same species.

This variety has not yet been recorded from other districts of Gondwanaland except in India. It is for the first time discovered in Africa.

Locality: The Idusi River, Ketewaka-Mchuchuma Coalfield.

Horizon: The Siltstone of the Idusi Series.

Gangamopteris cyclopteroides cf. var. *attenuata* Feistm.

Pl. I, fig. 6.

This is a complete specimen of a small leaf. The leaf is ligulate, 6cm. long and 1.8cm. broad in the widest, entire, with a bluntly pointed apex and a very narrow base. The median region is occupied by a group of grosser subparallel veins somewhat elevated,especially in the lower portion of the leaf. From those emerge the slightly arched secondary veins which are dichotomous and anastomose to form network.

In regard with the small ligulate shape and the narrow base as well as somewhat elevated madian subparallel veins, the present specimen is closely similar to *Gangamopteris cyclopteroides* var. *attenuata* (Feistmentel, O.1879, p.14.Pl.XIII, fig. 3:Pl.XVI, fig.5) from the Karharbari Stage in the Karharbari Coalfield of India.

Locality: The Mchuchuma River, Ketewaka-Mchuchuma Coalfield.

Horizon: The Siltstone of the Idusi Series.

Gangamopteris cf. *fibrosa* Maithy

Pl.I.figs.3,3a; Pl. II ,fig.1; Pl.III ,figs.1,1a; Pl.IV ,fig.1.

The material treated consists of a few specimens, of which the four well -preserved ones are shown in the plates and described below:

The leaf is lanceolate, attaining a length of 18.4cm. and breadth of 6.2 cm., entire, with narrowly obtuse apex. There is a well-marked shallow and broad furrow in the median which runs from the base to the apex, bears the fine crowded longitudinal striae. The secondary veins arise from the ɪedian furrow at a very acute angle,sometimes give off from the stem,gradually bend outwards to the edge, bifurcate, join to form the meshes which are broader and shorter near the median region.The veins towards the centre number 20-24 to a centimetre,30-34 near the margin.In the lamina there are very much dense and fine fibres, drawing out along the veins and numbering 7-9 to a millimetre.

The specimen shown in Pl. II , fig.1 is the largest. On the impression it appear that there are some veins giving off directly from the stem. The leaf impression shown in Pl. IV ,fig.1 is destroyed at the extreme base. On each of the specimens mentioned above, it can be seen that there is a remarkable furrow persisted to the top closely resembling a midrib. On the impression shown in Pl.I, fig.3 the median furrow is less remarkable as compared with the specimens mentioned above and it also almost persists to the top. The specimen in P. Ⅲ, fig.1 represents the leaf base, where it can be seen that there are some veins giving off directly from the stem. Pl.I, fig. 3a is an nlargement of the specimen shown in Pl. I, fig. 3 to show the details of the fibres on the lamina.

The specimens of this species display a quite variable degree of the shape and the median furrow of the leaf but they all possess very uniform venation and the characteristic fibres.

The most prominent feature of these specimens is the possession of fibres.Up to the present, in the genus *Gangamopteris* only *Gangamopteris fibrosa* which was recorded in the Karharbari Stage of Giridin Coalfield in India by Maithy (1964, Pl.4,figs.26,27,28) is known with the fibres. In accordance with the venation and fibres the present specimens agree perfectly with the Maithy's. However, one point which may be thought to render thei reference to this species somewhat doubtful is the still more apparent furrow.This difference, however, can not be regarded as a sufficient importance for specific separation. It should be noted that the Maithy's specimen in Pl.Ⅳ, fig. 27 represents only lower portion of a small leaf and the shape of its upper portion is unknown. Again, Maithy's specimen in Pl. Ⅳ,fig.26 is with a narrower leaf, in which the median region is occupied only by few subparallel veins which, however, do not form a distinct furrow. This specimen is considerably different from the Tanzania's specimens. However, it would be difficult in that case to decide that the Tanzania's specimens certainly belong to the Indian species. At present, all the specimens may be compared, as a provisional measure, with *Gangamopteris fibrosa*.

The form of this species is intermediate between *Gangamopteris* and *Glossopteris*. There is, on the leaf, a shallow and broad median furrow which is very closely similar to a midrib. On the leaf base some veins are given off directly from the stem.

Locality: The Mchuchuma River, Ketewaka-Mchuchuma Coalfield.

Horizon: The Siltstone of the Idusi Series.

Gangamopteris? sp.(sp. nov.?)

Pl.Ⅳ, figs. 4,4a

The small impression represents the upper portion of the leaf and it measures a length of 3.7cm. and a breadth of 2.7cm., with a rounded apex.The veins give off from the median subparallel veins at a very acute angle, gradually bend towards the edge, bifurcate, anastomose to form network. On the impression innumerable irregular minute pits are scattered.

The generic determination of the specimen is in some degree doubtful because it is only one of the upper portion of the leaf and it is yet unknown that whether there is a midrib in its middle and lower portion.By having many characteristic minute pits this specimen differs from any *Gangamopteris* and *Glossopteris* previously known. It perhaps represents a new species.

Locality: The Mbamba-Bay Coalfield.

Horizon: The Siltstone of the Idusi Series.

Gangamopteris sp. A.

Pl. V, fig. 1.

This fragmentary impression represents the upper portion of the leaf and it measures 5.8cm. long and 3.5cm. broad, and bears rounded apex. The midrib is absent.The veins radiate out from the base,dichotomous,anastomose to form the meshes which are almost uniform on the lamina. By the denser and more uniform meshes this specimen can be distinguished from the other materials.

Locality: The Mchuchuma River Ketewaka-Metewaka-Mchuchuma Coalfield.

Horizon: The Siltstone of the Idusi Series.

Gangamopteris sp. B.

Pl. II, fig. 3.

It is a specimen in bad preservation. The leaf is spatulate, 6.2cm. long and 2.4cm. broad, with rounded top, entire, gradually narrowing downwards. The slightly arched secondary veins arise from the median region of the leaf at a very acute angle, where they, however, do not join to from a group of the subparallel veins, as is seen in other species of *Gangamopteris*, e.g. *G. cyclopteroides*.

The specimen differs from *Gangamopteris* sp. A in the narrower and longer meshes.

Locality: The Mchuchuma River, Ketewaka-Mchuchuma Coalfield.

Horizon: The Siltstone of the Idusi Series.

Glossopteris indica Schimper

Pl. II, fig. 4.

In the present material,this species is represented by only one specimen in bad preservation, with two impressions broken off at the upper parts of the leaves. The left one is larger, 7.4cm. long and 2.6cm. broad, gradually narrowing towards the base into a petiole.The margin is entire.The midrib is distinct but does not exceed 2 mm. in the breadth. The secondary veins arise

from the midrib at an acute angle,bend outwards,bifurcate,anastomose to from the meshes which are broader and somewhat shorter towards the midrib. The veins near the midrib number about 24 to a centimetre and near the margin about 35. According to the leaf shape and venation the right one is in complete conormity with the one mentioned above. It may be seen that there are lumps on the impression of the midrib.

In regard to the narrow shape of the leaf and the broader and shorter meshes near the midrib the Tanzania's specimen is in perfect agreement with *Glossopteris indica* (Feistmantel, 0. 1880,p.101, Pl.38A, fig.4) from the Barakar Group of the Indian Talchir

Coalfield, but the only difference is that the veins of the former are more curved near the midrib. This difference,however, cannot be regarded as a great importance. It is not much doubt that this specimen would be referred to *Glossopteris indica.*

Locality: The Nyakalai River, Ketewaka-Metewaka-Mchuchuma Coalfield.

Horizon: The Mhukuru Series.

Glossopteris browniana Brongn

Pl.Ⅲ,figs.2,2a; Pl.Ⅳ,figs.2-3; Pl.V.fig.4.

The specimens of this species are few in number and can not sufficiently display the shape and size of the leaf.The specimen in Pl.V,fig.4 represents the upper part of leaf which measures 9.5 cm.long and 5.5 cm.broad,and bears the rounded apex. The midrib is distinct, 3mm. broad, with the fine longitudinal striae. The secondary veins give off from the midrib at a quite acute angle, abruptly bend outwards and pursue a straight course ôo the edge at an angle of about 80° ,bifurcate, anastomose with each other into meshes which are relatively uniform on all lamina. The veins near the midrib number about 22 to a centimetre, about 26 near the margin.

The specimen shown in Pl. Ⅲ, fig.2 represents the lower part of a leaf, which measures 9cm. long and 3.4cm. broad and attenuate downwards into a petiole. The midrib is distinct, with minute apparent lumps. The number of the veins towards the midrib is about 16 to a centimetre. Pl.Ⅲ,fig.2a is an enlargement of a part of the specimen shown in Pl.Ⅲ, fig2 for showing the details of the venation.In possession of the lumps and loose veins this specimen is little different from the specimen mentioned above.

It is stated that in *Glossopteris* the lumps, which probably are the internal structure exposed in preservation, are occasionally seen. Both the striae and lumps are seen simutaneously in one species, it appears that the lumps are worthless in a diagnosis.

The present specimens agree in most respects with *Glossopteris browniana* (Feistmantel, 0. 1878,p.78, Pl.Ⅷ, fig.4; Pl.X,figs.1,3; Pl.Ⅷ(XXVI), fig.1) from Australia but the former somewhat differs from the latter in the rather dense veins.Of *Glossopteris browniana* the venation has a considerable variation. For example, the two specimens from Transvaal have been referred to *Glossopteris browniana* by Zeiller (1896, p.362, Pl.XVI,figs.4,4A, 5,5A,5B.), of which the veins are denser in comparison with the Feistmantel's but agree with the specimens from Tanzania.

Locality: The Mhukuru River, Mhukuru Coalfield.

Horizon: The Mhukuru Series.

Vertebraria sp.

Pl.Ⅵ, fig.3.

It is a small fragmentary specimen which represents an internal model of a medulla of Glossopteris, measuring 2.5cm.long and 0.7cm. broad, consisting of two longitudinal series of rectangular areas separated by a longitudinal groove.The areas of the longitudinal series are transversely separated again by other grooves.

Locality: The Mhukuru River, Mhukuru Coalfield.

Horizon: The Mhukuru Series.

Noeggerathiopsis hislopi (Bunbury) Feistm.

Pl. V,fig.3; Pl. VI ,figs.1-2.

In the present material, this species is represented by some fragmentary specimens without leaf apexes. The specimen shown in Pl. VI .fig.1 has two impressions destroyed at both ends. The right one is larger,attaining a length of 21.8cm.and breadth of 4 cm.in the upper half. The veins are approximately parallel, dichotomous, numbering about 18 in a centimetre towards the upper part of the leaf. On the lamina, there are slender dense strae running along the veins. The specimen shown in Pl. VI, fig. 2 displays the lower part of a leaf measuring a length of 16.3cm.and a breadth of 14cm.,gradually narrowing towards the base.Radiating from the base,the veins are rather sparse as compared with the specimen mentioned above. As shown in Pl.V,fig.3 the specimen is well-preserved on the thin brownish grey shale and indicates the relatively distinct venation.

Thses specimens agree in every respect with *Noeggerathiopis hislopi* (Feistmantel,0. 1880,Pl.XXVIII, figs.1-2,4,6; Pl.XXIX, figs.1-4) from the Talchir Group in Deogher and Kara Coalfields and Karharbari Stage in the Karharbari Coalfield in India.After careful comparison with Feistmantel's figures, it is difficult to find any difference between them.

Noeggerathiopsis was created by Feistmantel (1878) for *Noeggerathis*? (*Cyclopteris* ?) *hislopi* which was published by Bunbury (1861) based on specimens from the Indian Nagpur. He gave to his species a wide range.According to the epidermic tissue and the form of the leaf, Lele & Maithy (1963) subdivided Nocggerathiopsis hislopi into three species, i.e.*Noeggerathiopsis indica, N. gondwanensis,N.zeiller*. The present specimens consist of impressions with no observable epidermis,therefore they would be changed to as *Noeggera thiopsis*.

Locality: The Idusi River, Ketewaka-Mchuchuma Coalfield.

Horizon: The Siltstone of the Idusi Series.

Palaeovittaria tanzaniensis Hu et He sp. nov.

Pl. V. fig. 2.

Leaf spatulate, entire, at least 12.4cm. long, 4cm.broad, gradually narrowing towards the base. Midrib scarcely visible, only indicated by a median fold in the lower portion, and dissolving upwards into fine veins. Secondary veins arising at a quite acute angle of about 5 from the midrib, straightly running to the apex and the margin, simple or bifurcating once or twice.Number of the veins uniform, about of 22 to a centimetre.

The material on which this new species is founded is represented only by one specimen in bad preservation broken at both ends.

The chief feature of this specimen is the scarcely visible midrib indicated only by a median fold in the lower portion of the leaf.This feature is in perfect conformity with the genotype of *Palaeovittaria* created by Feistmantel (1881).

Up to the present, in the genus *Palaeovittaria*, only two species *Palaeovittaria kurzi* and *P. mccarthyi* have been described and the records are very few.Owing to the much smaller angle formed by secondary veins and the midrib and the approximately parallel and rather dense veins the present specimen is considerably different from *Palaeovittaria kurzi*. It is very different from *Palaeovittaria mccarthyi* not only in the much smaller angle but especially in the straight veins.

The specimen from the Lukugo Series of the Gongo was referred by Hoeg and Bose (1960, XXXII,8,p.58, Pl.14,figs.7-8) to *Palaeovittaria*.In accordance with the denseness of the veins and the angle formed by the secondary veins and the median fold, it is a very close resemblance to the Tanzania's specimens. However, in the former the veins is more apparently curved inwards.Under the circumstance, it is uncertain that they belong to the same species.

CORRELATION AND DISCUSSION

The lower part of the Idusi Series of Tanzania consists of pebble-sands-tones, while the middle part is mainly composed of well-laminated, paper-like green mudstones which usually exhibit a characteristic rhyme that indicates a glacial or periglacial aspect. These have been considered to be the deposits belonging to the Upper Carboniferous Dwyka Tillite of South Africa (S-pence 1952; Mazurczak 1953). The Lower Mchuchuma which is a series of sand-stones and shales with coal seams, overlying the Idusi Series, has been considered to be compared with the Lower Permian Ecca Series of South Africa. Both the Idusi and the Lower Mchuchuma Series have altogether been named as the Karroo System (Stockley and Oates 1931; McKinlay 1954; Furon 1963). But according to some authors (Stockley and Oates 1931; McKinlay 1964),the upper part of the Idusi Series, the Siltstone, considered as the bottom layer of the coal measures,was assigned to the Early Permian. Investigating the sporo-measures overlying it (K2e),Hart (1960,1963,1965) recognised that the sporo-pollen assemblages of these two series are very different from each other, but he still followed the classification of previous authors and assigned both the Siltstone and the coal-measures to the Early Permian.

The older beds of the Lukuga Series of the Congo, the west neighbour of Tanzania, are glacial and periglacial deposits. They are overlain by black shales and coal-measures which were considered to be the equivalents of the Dwyka and the Ecca Series of the Lower Karroo in South Africa.Their geological age was assigned to the Late

Carboniferous and Early Permian.In 1954 Cahen assigned the black shales (W4) of Walikale of the Congo to the Carboniferous, but in 1961 he changed his viewpoint and assumed that the geological age of the glacial and periglacial shales of the Elila River of the Congo is Carboniferous or? Permo-Carboniferous and that of the black shales of Waikale and Lukuga is Permian. In 1960, having investigated the *Glossopteris* flora of the Congo, Hoeg and Bose assigned the black shales of Walikale to Permian-Carboniferous, and the black shales and the coalmeasures of Lukuga to Permian. In 1966, basing upon the characteristics of the sporopollen assemblages, Bose and Kar suggested that the geological age of the glacial and periglacial shales of the Elila River of the Congo is corresponding to that of the Talchir Series of India and the geological age of the black shales of Walikale is apparoximatly corresponding to a period between the Rikba stage of the Talchir and the Lower Barakar.

Hence, it is generally accepted that the coal measures overlying the Siltstone belong to the Lower Permian and the underlying beds corresponding to the glacial tillite belong to the Upper Carboniferous. The question is therefore concentrated on the part of the Siltstones or its equivalent. Should it be grouped with the upper beds or with the lower beds? Till now, some authors assign it to the Upper Carboniferous, while others to the Lower Permian, and the rest to Permo-Carboniferous, as a transitional stage from the Upper Carboniferous to the Lower Permian. Owing to lack of sufficient evidence, the question is still open to discussion.

From the viewpoint of sedimentation, it may be explained that the Sandstone belongs to a kind of tillite with sandy facies deposited in a basin, the Mudstone belongs to a kind of glacial lake deposit with lake mudstone facies, and up to the Siltstone,the sediments turn to be sandy clayey lacustrine bog deposits. Although each of the three stages possesses its own special features,they are closely related one another and all are the products of cold or cooler environments.The appearance of the "Camp" Sandstone with deltaic facies means that there must have been a big change in sediment environment. And the coal-measures indicate that the environment at that time was warm and wet.Thus,the sediment environments of the lower and the upper series are quite different. The disconformity between two series found in the Mchuchuma River indicates that there was a sedimental break between them. Furthermore, the Lower Mchuchuma Series sometime covers directly on the old strata. Therefore, the Siltstone and the Lower Mchuchuma Series are apparently not deposited in continuous sequence.

Judging from the floristic components, the flora of the Siltstone is comparatively simple. It is characterised by the predominance of *Gangamopteris*, and the complete absence of the genus *Glossopteris*.It is certain that it belongs to the early stage of the *Gangamopteris-Glossopteris* Flora. *Gangamopteris* spp., except *G.* cf. *fibrosa* Maithy and *G.cyclopteroides* cf. var. *attenuata* Feistm., are usually found in the Upper Carboniferous beds of Gondwanaland. Therefore, this flora looks very much like the flora of the

Talchirs of India. *Gangamopteris cyclopteroides* cf. var. *attenuata* Feistm. has been found in the Lower Permian Karharbari Stage of the Karharbari Coalfield, India. *Gangamopteris fibrosa* Maithy, a rare fossil, is only recorded once from the Karharbari Stage of Giridin Coalfield, India. Morphologically this species is more or less close to the genus *Glossopteris. Palaeovittaria tanzaiensis*, a species described as new in this work, and two formally known species of *Palaeovittaria* (*P. kurzi* Feistm. and *P. mccarthyi* Carey) occur in the Permian beds of Gondwanaland. Hence, the flora of the Sillstone bears the transitional characters between Carboniferous and Permian. It is noted that the genus *Glossopteris* is usually found in the Permian beds of Gondwanaland, but not a single specimen of this genus has been obtained from the Siltstone.

Up to the present,the flora of the Dwyka Series of South Africa has been insufficiently investigated and this makes great difficulties in comparison. Basing on the present materials,we know that the Lower Shale of South Africa contains *Phyllotheca*? *sp.* and *Cyclodendron*,the glacial beds yield *Gangamopteris cyclopteroides* Feistm., the Upper Shale has *Lepidodendron australe* M'coy (specific determination doubtful) and *Dadoxylon sp.*, and the "White Band" overlying it contains *Glossopteris sp.* It is generally recognised that the geological age of the Dwyka Series is Late Carboniferous and the Upper Shale is "not younger than Lowest Permian" (Du Toit 1956, page 356). The geological age of the "White Band" is generally considered to be Lower Permian. Judging from the present evidence, one may find that the Siltstone of the Idusi Series of Tanzania is quite close to the Dwyka of South Africa. Both of them are characterized similarly by the predominance of *Gangamopteris cyclopteroides* Feistm. and the complete absence of the genus *Glossopteris*.

Hoeg and Bose (1960) recorded some fossil plants, from the black shales (W4) of the Walikale region of the Congo, namely *Gangamopteris cyclopteroides* Feistm., *Noeggerathlopsis hislopi* (Bunb.) Feistm., a cone, foliage shoots, leaves, some stem impressions and incrustations, *Samaropsis boutakoffii* Hoeg et Bose, *Cordaicarpus sp.*, and *Walikalia chaensii* Hoeg et Bose. Among which, *Gangamopteris cyclopteroides* is extremely common. The last three are fossil seeds and scale leaves (?) which are very rare and only one or few specimens to each species have been obtained. Although few fossils occur in the Walikale region, the predominance of *Gangamopteris cyclopteroides* Feistm.,and the complete absence of the genus *Glossopteris* are remarkable and are the characteristics both of the Tanzania and the Congo floras. Hence, it can be recognized that their geological ages are rather near.

The flora of the Siltstone also resembles that of the Talchirs of India. All its fossils, except *Palaeovittaria tanzaniensis* and *Gangamopteris* cf.*fibrosa* Maithy, are also very common in the Talchirs. The fact that *Glossopteris* has been recorded from the Talchirs found in Rikba (Krishnan 1956) and South Rewah (Fox 1931 p.220) of India indicates that the Siltstone is probably synchronous with the Talchirs, if not slightly earlier than the later.

Generally speaking, *Gangamopteris* appears to be the early representative of the *Gangamopteris-Glossopteris* flora, and it is found even in the intercalated glacial beds.But *Glossopteris* mostly occurs in the transitional beds of Permo-Carboniferous and in the Permian.

Although the genus *Palaeovittaria* somewhat with a Permian aspect occurs individually, yet in the Siltstone, the main elements are *Gangamopteris*, and no *Glossopteris* has so far been observed, while in the beds above the "Camp" Sandstone, *Gangamopteris* is present, but *Glossopteris* is predominant,showing the typical characteristics of the "karroo Series" . We have found *Glossopteris* browniana Brongn., *G. indica* Feistm., and *Vertebraria* sp. in the Mhukuru Series, but no *Gangamopteris* has been observed. Pant (1958) has also studied the flora of the Mhukuru, and no *Gangamopteris* has been recorded by him, too.

CONCLUSION

1.The present paper describes in detail the characteristics of the plant fossils represented by *Gangamopteris* flora found in the Siltstone Stage in southwestern Tanzania. It also describes the characteristics of the new fossil plant—*Palaeovittaria tanzaniensis*.

2.From the sedimentation point of view, the Sandstone Stage in the lower part of Idusi Series belongs to glacial silty conglomerate; the Mudstone Stage possesses typical varves and can be regarded as glacial-lake deposits; the Siltstone Stage is characterized by an abundance of plant fossils, it has a steady lithological character and belongs to lacustrine bog deposits, all the shows that the climate at that time had gradually warmed up, but compared with the overlying coal-formations, it was still a cold climate.Therefore, the Siltstone Stage has a transitional nature and can even be regarded as a transitional stage from the Upper Carboniferous to the Lower Permian.

3.Judging from the fossil flora, it can be seen that the fossil plants in the Siltstone Stage mainly are the elements of Carboniferous age, although they preserve some elements of the Permian age.

4.Taking into consideration the establishment of straugraphical system of geological formations and convenience for exploitation in Tanzania and the fact that Lower Mchuchuma Series contains coal seams of economic value while the Idusi Series does not, it is more appropriate to take the transitional Siltstone Stage temporarily as belonging to the upper part of Idusi Series and to the Upper Carboniferous. It also describes the characteristics of two main elements of Gondwana flora-*Glossopterls in dica* schimper and *G.browniana* Brongn. from Mhukuru Series, Upper Permian.

REFERENCES

[1] Arber, E.A.N.,1902. On the Clarke collection of fossil plants from New South Wales. *Quart.J.Geol. Soc. Lond.* 58:1-26.

[2] Arber, E. A.N.,1905. Catalogue of the fossil plants of the *Glossopteris* flora in the Department of Geology. *British Museum (Natural History) London.*

[3] Balme, B.E.,1952. On some spore specimens srom British Upper Carboniferous coals. *Geol. Mag.* 89(1-6): 175-184.

[4] Balme, B.E. & J.P.F. Hennelly, 1955. Bisaccate sporomorphs from Australian Permian coals. *Austral. J.Bot.* 3:89-98.

[5] Balme,B.E.,1956a.Monolete,Monocolpate and Alete sporomorphs from Australian Permian Sediments. Ibid. 4:54-67.

[6] Balme,B.E.,1956b. Trilete sporomorphs from Australian Permian Sediments. Ibid. 4:240-260.

[7] Beug,H.J.,1956. Pollendimorphisums bei *Ephedra. Naturwissuschaftes 43:332*-333.

[8] Bharadwaj, D.C.,1955. The spore genera from the Upper Carboniferous Coals of the Saar and their value in Stratigraphical Studies. *Palacobotanist* 4:119-149.

[9] Bharadwaj, D.C.,1957. The spore flora of Velener Schichten (Lower Westphanian D) in the Ruhr Coal Measurcs. *Palacontographica* B, 102:110-138.

[10] Bharadwaj, D.C.,1962. The miospore genera in the coals of Ranigani Stage (Upper Permian), India. *Palacobotanist* 9:68-106.

[11] Bharadwaj, D.C. & G. Kremp, 1955. Dic Sporenfuhrung der Velener Schichten des Ruhrkarbons. *Geol.Jb.* Band. 71.

[12] Bharadwaj, D.C. & S. K. Salujha,1964. Sporological study of seam Ⅷ in Raniganj Coalfield,Bihar (India), part 1.-Description of sporae disperase. *Palacobotanist* 12 (2):181-215.

[13] Bharadwaj, D.C. & R.S.Tiwari, 1964. On two monosaccate genera from Barakar Stage of India. *Palacobotanist* 12(2):139-146.

[14] Bharadwaj, D.C. & B.S.Venkatachala, 1957. Microfloristic evidence on the boundary between the Carboniferous and the Permian System in Pfatz (W. Germany). *Palacobotanist* 5(1):1-11.

[15] Bhattacharyya, A.K., 1963. The assemblage of megaplant fossil from the Lower Gondwana rocks of the western part of the Auranga Valley Coalfield, Palamau district, Bihar. *Quart.J.Geol.Soc.* 35:123-128.

[16] Bose,M.N. & R.K. Kar, 1966-67. Palaeozoic spore dispersae from Gongo.*Ann.* Mus. Roy. Afr. Centr., Tervuren, in-8 Geol.

[17] Bunbury,C.J.F.,1861. Notes on a collection of fossil plants from Nagpur, Central India. *Quart. J. Geol.Soc.* 7:325-340.

[18] Du Toit,A.L.,1956. The geology of South Africa. Reprinted.

[19] Feistmantel,O.,1876a,On some fossil plants from the Damuda Series in the Raniganj Coalfield,collected by Mr.J.Wood-Mason.*J.Asiat.Soc.Beng*.45:239.

[20] Feistmantel,O.,1876b. Notes on the age of some fossil of India. *Rec.Geol. Surv. India* 9(3):63-79.

[21] Feistmantel, O., 1879-80. The fossil flora of the Lower Gondwanas 1.The flora of the Talchir Karharbari beds.*Palaeont.Indica*. Ser.Ⅶ,3(1):1-64.

[22] Feistmantel,O.,1880-1881. The fossil flora of the Gondwana System (Lower Gondwana) 2.The flora of Danuda-Panchet Division. *Palaeont.Indica*. Ser.Ⅶ,3(2):1-146.

[23] Feistmantel, O., 1882. The fossil flora of the South Rewa Gondwana Basin. Ibid. Ⅶ, 4(1):1-52.

[24] Feistmantel,O.,1886. The fossil flora of the Gondwana System, 4.The fossil flora of some coalfields in Western Bengal.Ibid. Ⅶ,4(2):1-66.

[25] Feistmantel, O., 1890. Geological and palaeonotological relations of the coal and plant-bearing beds of Palaeozoic and Mesozoic age in Eastern Australia and Tasmania,etc. *Mem.Geol.Surv.N.S.* 11:(*Palaeont*) 3:1-183.

[26] Fox,C.S.,1934.The Lower Gondwana Coalfields of India.*Nem.Geol.Surv.India* 59:78-85.

[27] Furon, R.,1963. Geology of Africa.

[28] Ghosh,A.K. & J. Sen, 1948. A study of microfossils and the correlation of some productivecoal seam of the Raniganj Coalfield,Bengal,India.*Trans. Min.Geol.Inst.India*. 43:67-93.

[29] Harkin,D.A.,1948.The Mbamba-Bay Coalfield,Tanganyika Territory. *Min.Mag.* 78:265-272.(Reprints issued as *Min.Res.Pamphal.Geol.Sury.Tanganyika*,49).

[30] Harkin,D.A.,1953. The geology of the Mhukuru Coalfield, Songea district. *Short Pap. Geol.Surv.Tanganyika*. 28.

[31] Harkin,D.A.,1955. The geology of the Songwe-Kiwira Coalfield,Rungwe district. *Bull. Geol.Surv.Tanganyika*,27.

[32] Hart,G.F.,1960.Microfloral investigation of the Lower Coal Measures (K2): Ketewaka-Mchuchuma-Coalfield,Tanganyika. *Geol. Surv.Tang.13:1-18*.

[33] Hart.G.F.,1962. Palynology-The key to stratigraphy? *S.Afr.J. Sci*.58(12): 365-374.

[34] Hart,G.F.,1963.Microfloral investigation of the Lower Coal Measures (K1): Ketewaka-Mchuchuma-Coalfield. Tanganyika.

[35] Hart, G.F.,1963. A probable pre-*Glossopteris* microfloral assemblage from Lower Karroo sediments. *S.Afr.J.Sci*. 59(5):135-146.

[36] Hart. G.F.,1964. A review of the classification anddistribution of the Permaian miospore:Disaccate stratiti.

[37] Hart,G.F.,1965. Miospore zone in Karroo sediment of Tanzania. *Palae.Afr*. 9:139-150. Illus.

[38] Hoeg,O.A. & M.N.Bose,1960. The *Glossopertis* flora of the Belgian Congo with a note on some fossil plants from the Zambesi Basin (Mozambiquc). *Ann . Mus.Roy.Congo Belge*, 8:32-99.

[39] Hoeg,O.A.,1961.The *Glossopteris* flora of the Belgian Congo. *Ann. du Mus. Royal du CongoBelge* 8:*Sci.geol.* 32.

[40] Kosanke,R.M.,1950.Pennsulvanian spores of Illinois and their use in correlation. *State Geol.Surv.* Ill.*Bull.* 74:1-128.

[41] Krishnan, M.S.,1956. Geology of India and Burma. Reprinted.

[42] Lele,K.M.& P.K. Maithy,1963. Studies in the *Glossopteris* flora of India-15.Revision of theepidermal structure of *Noeggerathiopsis* Feistmantel. *Palacobotanist.* 12(1):7-17.

[43] Leschik,G.1959.,Sporen aus den "Karru-Sandsternen" von Norronaud (Sudwest Afrika). *Senekenbergiana* 40:52-59.

[44] Maheshwari, H.K.,1965a. Studies in the *Glossopteris* flora of India-22. On some species of thegenus *Glossopteris* from the Raniganj Stage of east Raniganj Coalfield, Bengal. *Palaeobotanist* 13 (2):129-140.

[45] Maheshwari,H.K.1956b.Studies in the *Glossopteris* flora of Indica-23. On two fructifications from the Raniganj Stage of the Raniganj Coalfield, Bengal. *Palacobotanist* 13(2).

[46] Maheshwari,H.K. & G. Prakash, 1965. Studies in the *Glossopteris* flora of India-21. Plant megafossils from the Lower Gondwana exposares along Ban sloi river in Rajmahal Hills, Bihar. *Palaeobotanist* 13(2):115-128.

[47] Maithy,P.K.,1965. Studies in the *Glossopteris* flora of India-26. Glossopteridales from the Karharbari beds. Giridih Coalfield,India.Ibid.13(3): 248-262.

[48] McCoy, F.,1847. On the fossil botany and zoology of the rocks associated with the coal of Australia. *Ann. Mag. Nat. Hist.*(1) 20:151.

[49] McKinlay, A.C.M.,1954. Geology of the Ketewaka-Mchuchuma-Coaifield, Njom-be District. *Bull.Geol.Surv.Tanganyika* 21.

[50] McKinlay,A.C.M.,1965. The coalfields and the coal resources of Tanzania. *Bull.Geol.Surv.anzania* 38.

[51] Moore,L.R.,1960. Summary report on the Ketewaka-Mchuchuma Coalfield.*Rec. Geol.Surv.Tanganyika* 8:19-58,54-62.

[52] Pant, D. D.,1949. On the occurrence of *Pityosporites* Seward in a Lower Gondwana Tillite from Australia and its possible relationships with *Glossopteris*. *Proc.*36th.,*Indian Sci.Congr.Allahd.*4:10-11.

[53] Pant, D.D.,1955. On two disaccate spores from the Bacchus Marsh Tillite, Victoria (Australia). *Ann.Mag.Nat.Hist.* 8:757-764.

[54] Pant,D.D., 1958. The structure of some leaves and fructifications of the *Glossopteris-flora* of Tanganyika. *Bull.Brit.Mus.(Nat.Hist.)* 3:127-176.

[55] Pant,D.D.,1960. Some seed and sporangia of *Glossopteris* flora from Raniganj Coalfleld, India. *Palaeontographica* 107.

[56] Piumstead,E.P.,1952.Description of two new genera and six new species of furetifications borne on *Glossopteris* leaves. *Trans. Geol.Soc.S.Afr*.55: 281-328.

[57] Plumstead,E.P.,1956. Bisexual fructifications borne on *Glossopteris* leaves from South Africa. *Palaeontographica* 100B:1-25.

[58] Plumstead, E.P.,1956. On *Ottokaria*, the fructification of *Gangamopteris*. *Trans. Geol.Soc.S.Afr*. 59:211-326.

[59] Plumstead,E.P.,1958. Further fructifications of the *Glossopteridae* and a Provisional classification based on them. *Trans.Geol.Soc.S.Afr*.61:51-76.

[60] Potonie, R.,1956,1958,1960. Synopsis der Gattungen der Sporae dispersae, Teil 1-3. *Beih.Geol.Jb*. 23,31,39.

[61] Potonie, R., 1962. Synopsis der Sporae in Situ.*Beih.Geol.Jahr*. 52.

[62] Potonie, R.& Klaus,W.,1954. Einige Sporegattungen des alpinen Salzgebriges. *Geel.Jb*. 68:517-544.

[63] Potonic,R. & Kremp G.,1954. Die Gattungen der Palacozoischen Sporae dispersae und ihre Stratigraphic. *Geol.Jb*. 69:111-194.

[64] Potonic,R.,1955,1956.Die Sporae dispersae des Ruhrkarbons usw.I.*Palacon-tographica*. 98:1-136; Ⅱ.Ibid.99:85-191:Ⅲ.Ibid.100:65-121.

[65] Potonie,R.& Lele,K.M.,1959.Studies in the Talchir flora of India-1. Sporae dispersae from the Talchir beds of South Rewa Gondwana Basin. *Palacobotanist* 8(1-2):22-37.

[66] Potonie,R.& Sah,S.C.D., 1960. Sporae disperae of the *Lignies* from Gannanore Beach of the Malabar Coast of India. Ibid. 7(2):121-135.

[67] Quennell, A.M.& McKinlay, A.C.M.,1956. Summary of the geology of Tanganyika. *Memoir* 1:81-98.

[68] Saksena,S.D.,1963.On fossil flora of Ganjra Nalla bed,South Rewa:part 1. Macrofossils. *Palaeobotanist* 11(1-2):23-29.

[69] Schemel,M.P.,1951. Small spores of Mystic Coal of Iowa.*Amer.Midl.Nat*. 46 (3):743-750.

[70] Sen,J.,1955.On some fructifications borne on *Glossopteris* leaves.*Bot.No*-tiser. 108:244-252.

[71] Seward, A.C., 1897. On the association of *Sigillaria* and *Glossopteris* in South Africa. *Quat.J.Geol.Soc*. 53,pl.318,t.21.f.4a.

[72] Seward,A.C.,1907.Fossil plants from South Africa. *Geol.Mag.N.S.Decade* V, 6:481-487.

[73] Seward, A. C., 1941. Plant life through the ages.Reprented.

[74] Seward, A.C.& Leslie, T.N.,1908. Permo-Carboniferous plant from Vereeniging. *Qourt.J.Geol.Soc.Lond*. 64:109-126.

[75] Seward, A.C.& Shani B.,1920.Indian Gondwana plants-A Revision. *Palaeont. Indica*. N.S.7(1):1-41.

[76] Spence,J.,1952.Ngaka Coalfield,Songea District. *Annu.Rep.Geol. Surv.Tanganyika*:18-21.

[77] Stockley,G.M.,1947.New Coal Discoveries in Tanganyika and coal resources of East and Central Africa.*Min.Mag*. 76(6):329-38; 77(1):9-16.

[78] Stockley,G.M.& Oates, F.,1931. Report on the geology of the Ruhuhu Coal-fields, Njombe-Songea Districts.*Bull.Geol.Surv*. Tanganyika Territory,2.

[79] Stockley,G.M.,1931. The Ruhuhu Coalfields,Tanganyika Territory. *Quart.J. Geol.Soc.,Lond.*, 88.

[80] Surange,K.R.& Lele,K.M.,1955. Studies in the *Glossopteris* flora of India -3. Plant fossil from the Talchir Needle Shale from Giridih Coalfield. *Palaeobotanist* 3:153-157.

[81] Surange, K.R., 1956. Studies in the *Glassopteris* flora of India-6. Plant fossils from the Talchir Beds of South Rewa Gondwana Basin. *Palaeobotanist* 5 (2):82-90.

[82] Surange,K.R.& Maheshwari,H.K.,1962. Studies in the *Glossopteris* flora of India-11. Some observations on *Vertebraria* from the Lower Gondwana of India. *Palacobotanist* 9(1-2):61-67.

[83] Surange,K.R.& Srivastava,P.N.,1956. Studies in the *Glossopteris* flora of India-5.Generic status of *Glossopteris,Gangamopteris* and *Palaeovittaria*. Ibid. 5(1):46-49.

[84] Tenison-Woods,Rev.J.E.,1883. On the fossil flora of the coal deposits of Australia. *Pro.Linn.Soc.N.S.W*.8(1):127.

[85] Virkki,C.,1937. On the occurrence of winged spores in the Lower Gondwana Rocks of India and Australia. *Proc.Ind.Acad.Sci*. 6:428-431.

[86] Virkki, C.,1939. On the occurrence of similar spores in a Lower Gondwana Glacial Tillite from Australia and Lower Gondwana Shales in India.*Proc.Ind.Acad.Sci*. 9(1):7-12.

[87] Walkom, A.B.,1921. On a specimen of *Noeggerathiopsis* from the Lower Coal Measures of New South Wales. *Proc.Linn.Soc.New South Wales*. 46(3).

[88] Walton,J.,1929.The fossil flora of the Karroo System in Wankie district, Southern Rhodesia. *Bull.Geol.Surv.S.Rhodesia*. 15:62-75.

[89] Wilson, L.R.,1963a. Morphological variation of *Thymospora psedothiessenii* (Kosanke) Wilson and Venkatachala. *Okl.Geol.Notes*.

[90] Wilson, L.R.,1963b. *Thymospora*,a new name for *Verrucososporites.Okl.Geol.Nttes* 23:75-79.

[91] Wilson,L.R.& B.S.,Venkatachala, 1963. A morphologic study and emendation of *Vesicaspora* Schemel.*Oklah.Geol.Notes Oklah.Geol.Surv*.23. (6)142-149.

[92] Wilson,L.R.,1966.Type species of *Cirratriadites* Wilson and Coe.*Okl.Geol.Notes* 26(2).

图版 I

图1. 木贼类茎杆
产地：基德华加一姆丘丘玛煤田姆丘丘玛河
地层：依多西组粉砂岩段

图2. 圆恒河羊齿
产地：姆邦巴贝煤田
地层：依多西组粉砂岩段

图3，3a. 似条纹恒河羊齿 图3a是图3标本局部放大，用以显示叶面上条纹的性状，×3
产地：基德华加一姆丘丘玛煤田姆丘丘玛河
地层：依多西组粉砂岩段

图4－5. 似圆恒河羊齿
产地：姆邦巴贝煤田
地层：依多西组粉砂岩段

图6. 楔形恒河羊齿
产地：基德华加一姆丘丘玛煤田姆丘丘玛河
地层：依多西组粉砂岩段

图版 II

图1. 似条纹恒河羊齿
产地：基德华加一姆丘丘玛煤田姆丘丘玛河
地层：依多西组粉砂岩段

图2. 渐尖恒河羊齿
产地：基德华加一姆丘丘玛煤田依多西河
地层：依多西组粉砂岩段

图3. 恒河羊齿乙种
产地：基德华加一姆丘丘玛煤田姆丘丘玛河
地层：依多西组粉砂岩段

图4. 印度舌羊齿
产地：基德华加一姆丘丘玛煤田尼亚克莱河
地层：姆呼库鲁组

图版 III

图1,1a. 似条纹恒河羊齿 图1稍稍缩小一点，图1a是图3标本局部放大,用以显示叶基部的叶脉
产地：基德华加一姆丘丘玛煤田姆丘丘玛河
地层：依多西组粉砂岩段

图2,2a. 均网舌羊齿. 图2a是图2的局部放大，用以显示叶脉性状，× 3

产地：姆呼库鲁煤田姆呼库鲁河
地层：姆呼库鲁组

图版 Ⅳ

图1. 似条纹恒河羊齿
产地：基德华加一姆丘丘玛煤田姆丘丘玛河
地层：依多西组粉砂岩段

图2,3. 均网舌羊齿．图2显示的叶脉稍密，或许属于布郎舌羊齿种
产地：姆呼库鲁煤田姆呼库鲁河
地层：姆呼库鲁组

图4,4a. 恒河羊齿种(?新种)．图4a是图4的局部放大，用以显示叶面上的细小凹坑
产地：姆邦巴贝煤田
地层：依多西组粉砂岩段

图版 Ⅴ

图1,1a. 恒河羊齿甲种,图1a是同一图版图1的标本局部放大，用以显示叶脉的性状，×3
产地：基德华加一姆丘丘玛煤田姆丘丘玛河
地层：依多西组粉砂岩段

图2. 坦桑古书带叶．(新种)
产地：基德华加一姆丘丘玛煤田依多西河
地层：依多西组粉砂岩段

图3. 辐脉苛达．(新组合)
产地：基德华加一姆丘丘玛煤田依多西河
地层：依多西组粉砂岩段

图4. 均网舌羊齿
产地：姆呼库鲁煤田姆呼库鲁河
地层：姆呼库鲁组

图版 Ⅵ

图1－2.辐脉苛达．(新组合)
产地：基德华加一姆丘丘玛煤田依多河
地层：依多西组粉砂岩段

图3. 脊椎木的1种，图影略微放大了一点
产地：姆呼库鲁煤田姆呼库鲁河
地层：姆呼库鲁组

PLATE I

Fig.1. Equisetaceaous stem.
Locality: The Mchuchuma River.Ketewaka-Mchuchuma Coaltield.
Horizon: The Siltstone of the Idusi Series.

Fig.2. *Gangamopteris cyclopteroides* Feistm.
Locality: The Mbamba-Bay Coalfield.
Horizon: The Siltstone of the Idusi Series.

Figs.3,3a. *Gangamopteris* cf.*fibrosa* Maithy. Fig.3a is an enlargement of the part of the specimen in fig.3 to shown the details of the fibres. × 3
Locality: The Mchuchuma River, Ketewaka-Mchuchuma Coalfield.
Horizon: The Siltstone of the Idusi Series.

Figs.4,5. *Gangamopteris* cf.*cyclopteroides* Feistm.
Locality: The Mbamba-Bay Coalfield.
Horizon: The Siltstone of the Idusi Series.

Fig.6. *Gangamopteris cyclopteroides* cf. var.*attenuata* Feistm.
Locality: The Mchuchuma River, Ketewaka-Mchuchuma Coalfield.
Horizon: The Siltstone of the Idusi Series.

PLATE II

Fig.1. *Gangamopteris* cf.*fibrosa* Maithy.
Lcocality: The Mchuchuma River, Ketewaka-Mchuchuma Coalfield.
Horizon: The Siltstone of the Idusi Series.

Fig.2. *Gangamopteris cyclopteroides* var. *acuminata* Feistm.
Locality: The Idusi River, Ketewaka-Mchuchuma Coalfield.
Horizon: The Siltstone of the Idusi Series.

Fig.3. *Gangamopteris* sp. B
Locality: The Mchuchuma River, Ketewa-Mchuchuma Coalfield.
Horizon: The Siltstone of the Idusi Series.

Fig.4. *Glossopteris indica* Schimper.
Locality: The Nyakalai River, Ketewaka-Mchuchuma Coalfield.
Horizon: The Mhukuru Series.

PLATE III

Figs.1,1a.*Gangamopteris* cf.*fibrosa* Maithy. Fig.1 is slightly reduced; fig.1a is an enlargement of the part of the specimen in fig.1 for showing the details of the veins of the leaf base. × 3.
Locality: The Mchuchuma River,Ketewaka-Mchuchuma Coalfield
Horizon: The Siltstone of the Idusi Series.

Figs.2,2a.*Glossopteris browniana* Brongn. Fig.2a is enlarged the part of the specimen in

fig. 2 to show the details of the venation.
Locality: The Mhukuru River,Mhukuru Coalfield.
Horizon: The Mhukuru Series.

PLATE Ⅳ

Fig.1. *Gangamopteris* cf.*fibrosa* Maithy.
Locality: The Mchuchuma River,Ketewaka-Mchuchuma Coalfield.
Horizon: The Siltstone of the Idusi Series.

Figs.?2,3. *Glossopteris browniana* Brongn. Fig. 2 shows the denser veins and probably belongs to the same species.
Locality: The Mhukuru Rever Mhukuru Coalifield.
Horizon: The Mhukurur Series

Figs.4,4a.*Gangamopteris* sp. (? sp.nov.).Fig,4a is an enlargement of the part of the specimen in fig.4 to show the details of the minute pits on the lamina. × 3.
Locality: The Mbamba-Bay Coalfield.
Horizon: The Siltstone of the Idusi Series.

PLATE Ⅴ

Figs.1,1a.*Gangamopteris* sp. A Fig.1a is enlarged the part of the specimen in fig. 1 to show the details of the venation. × 3.
Locality: The Mchuchuma River,Ketewaka-Mchuchuma Coalfield.
Horizon: The Siltstone of the Idusi Series.

Fig.2. *Palaeovittaria tonzaniensis* Hu et He., sp. nov.
Locality: The Idusi River, Ketewaka-Mchuchuma Coalfield.
Horizon: The Siltstone of the Idusi Series.

Fig.3. *Cordaites hislopi* (Feistm.) Hu, comb. nov.
Locality: The Idusi River,Ketewaka-Mchuchuma Coalfield.
Horizon: The Siltstone of the Idusi Series.

Fig.4. *Glossopteris browniana* Brongn.
Locality: The Mhukuru River, Mhukuru Coalfield.
Horizon: The Mhukuru Series.

PLATE Ⅵ

Figs.1,2. *Cordaites hislopi* (Feistm.) Hu, comb. nov.
Locality: The Idusi River, Ketewaka-Mchuchuma Coalfield.
Horizon: The Siltstone of the Idusi Series.

Fig.3. *Vertebraria* sp. slightly enlarged.
Locality: The Mhukuru River.Mhukuru Coalfield.
Horizon: The Mhukuru Series.

PLATE I

PLATE II

×3

2

2a ×3

1

PLATE III

PLATE IV

1

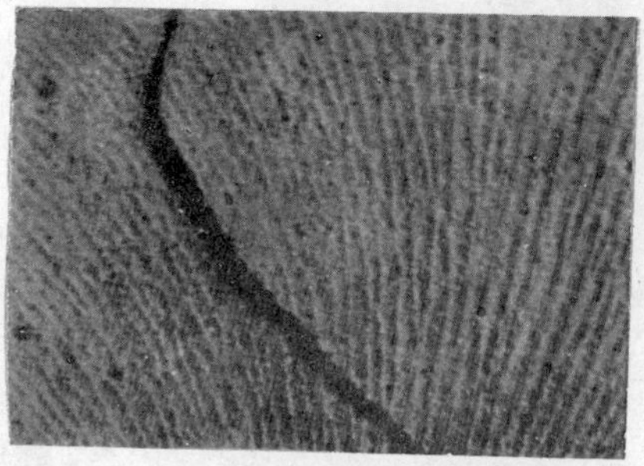

×3 1a

2

3

4

PLATE V

PLATE VI

中国濒危苔藓植物保护生物学研究

吴鹏程 贾渝 汪楣芝
(中国科学院植物研究所，北京 100093)

CONSERVATION BIOLOGY OF RARE AND ENDANGERED TAXA OF BRYOPHYTES IN CHINA

Wu Pan-cheng, Jia Yu and Wang Mei-zhi
(Institute of Botany, Academia Sinica, Beijing 100093, China)

Abstract: In China, the awareness of rare and endangered bryophytes was brought to attention when some 50 species of endangered Chinese bryophytes were listed by L. -Z. Chen et al. in 1993. Through field work and herbarium study, three steps of consideration to protect the rare and endangered species of bryophytes in China are proposed. They are: a).define the actual distribution of endangered species; b).investigate the biology, ecology, as well as genetics, of endangered species; and c).list the to-be-protected species.

The actual distribution and population size of rare and endangered bryophytes should be the first step to consider in determining the status of endangerment of a species and its classification into one of the five proposed categories: extinct, very endangered, endangered, declining,and not threatened. All selected species must be evaluated accordingly. A second area of investigation should focus on the ecology of endangered species and its biology, including its sexual and asexual reproductions. Finally, six species of endangered bryophytes and six regional reserves are selected for the conservation and protection of endangered bryophytes in China.

Key words: Rare and endangered species; Conservation; Bryophytes

摘要: 自1993年由陈灵芝等报道中国大约50个种的珍稀濒危苔藓植物后，此问题已引起人们的注意．通过野外和标本室研究,我们已考虑在中国有3方面工作可用以保护这些苔藓植物:包括1)濒危苔藓种的分布区调查 2)研究它们的生物学及生态学特性，如有可能必要时研究它们的遗传特性及 3)列出被保护的种类．

开展上述工作的第一个步骤，是研究和调查每一个被保护的苔藓物种的实际分布区和群丛大小,并将它们划分为5个等级:绝灭、十分濒危、濒危、略有减少及正常．在此基础上，对逐个种类的生物学特性包括有性与无性繁殖方式等的内在因素进行研究,以找出濒危原因．根据中国目前实际状况及条件等因素，建议对中国濒临减少的苔藓植物中选择6个种及6个地区加以保护开展研究．

关键词: 苔藓植物; 珍稀濒危种类; 保护生物学; 中国

Because of the changing landscape, air pollution and destruction of environment, species of bryophytes are decreasing in number in China. Others may become extinct soon. Recently,the awareness of the existence of rare and endangered taxa of bryophytes has gained public attention. About 50 taxa of Chinese bryophytes were listed by Z. -L. Chen et al. in 1993. Through field work and herbarium study, we suggest the undertaking of the following three steps designed to protect rare and endangered species in the country. They are: a).determine the actual range of rare and endangered bryophytes; b).investigate the ecology, biology and genetics of rare and endangered bryophytes; c).list the rare and endangered taxa of bryophytes that need protection.

1. Determination of distribution and population size of endangered taxa of bryophytes.

First, all the endemic Chinese bryophytes need to be protected. This is especially true for the East Asiatic endemics that are distributed only in China, Japan and Korea. They are important components of the Chinese flora. However, doubtfully described taxa should be excluded. In general,population size and distributional range should serve as the basis to determine the degree of endangerment of a species (see Table 1).

Table 1. **Proposed classification of rare and endangered bryophytes in China.**

Category	Distribution
extinct	recorded in China but no extant population
very endangered	found in 1-3 mountain regions
endangered	found in 3-5 mountain regions
declining	distributed in ca 3-5 provinces
non-threatened	widely distributed

Five categories are proposed for use in classifying the rare and endangered bryophytes in China. The foremost basis is the actual distribution of the species. Considering the physical size of the country, we propose the above stated standardized categories. For example, very endangered bryophytes are those that may occur in one or two provinces but not extending more than three mountain ranges. The size of each of the populations of an endangered species should be less than 3-5 sq. cm, at most less than 10 cm in diameter.

2. Investigation of the ecology and biology of rare and endangered species.

Investigation of rare and endangered bryophyte species should include the study of biology and habitat requirement of the species, such as:

A. Environmental factors. Like the vascular plants,the survival of bryophyte populations depend on environmental factors such as light, temperature and moisture. It is therefore important to observe the populations of endangered species in the same locality over the change of seasons and make comparison with the populations of other bryophyte species.

B. Biology. Study of the biology of rare and endangered species must concentrate on the effectiveness of asexual and sexual reproductions,especially the types of gemmae and other propagules. The sexual condition and location of sex organs on an individual plant are other important information needed for the effective preservation of rare and endangered species.

C. Genetics. Knowledge of genome is important because genetic characters are a complex issue among biological characteristics of rare and endangered bryophyte species. Analysis of DNA sequence is one method used to interpret the genetic heterogeneity of a population and may provide clue to help save a species.

3. Selection of bryophyte species and their areas/habitats for protection.

If we want to save rare and endangered species of bryophytes, we must preserve the habitats and leave the areas undisturbed.Considering the financial situation, one can not save all the endangered species of bryophytes. Probably only a few can be protected immediately. We therefore propose the followings:

A. Protected species - Considering the range information of all rare and endangered species in China, we recommend here two liverworts and four mosses for immediate protection. They are *Neotrichocolea bissetii, Pleurozia giganteoides, Pseudopterobryum laticuspis, Miyabea rotundifolia, Sciaromiopsis sinensis* and *Takakia ceratophylla* (see also List 1).

B. Regional preserves-Many rare and endangered tropical and subtropical species are found in southern and eastern parts of China. We propose Mengla County of Xishuangbanna, Mt. Jinfu in Sichuan Province, Mt. Diaolo in Hainan Island, Shennongjia Mt. and West Tianmu Mt. in eastern China,and Changbai Mt. in northeastern China as the six top areas for the preservation of rare and endangered bryophytes in China (Map 1).

Finally we should let everyone know that we have only one world. We must save as many rare and endangered species of organisms as possible. In doing so, we are also saving the human species.

Acknowledgements

The authors are deeply indebted to Dr. David Boufford of the Harvard University to help us to correct this paper carefully.

References

[1] Chen, L. Z. (Editor in Chief) et al., 1993. The Biodiversity of China, its present status and conservation policy. Science Press, Beijing.

[2] Jia, Y. et P. C. Wu, 1995. The rare and endangered bryophytes in China and their conservation. In Qian, Y. Q. (Editor in Chief), Advances in Biodiversity Research, P.215-220. Chinese Science and Technology Press. Beijing.

List 1. A list of the rare and endangered taxa of bryophytes in China.

Takakiaceae
Takakia ceratophylla
Takakia lepidozioides

Trichocoleaceae
Neotrichocolea bissetii
Trichocoleopsis sacculata
Trichocoleopsis tsinlingensis

Lepidoziaceae
Zoopsis liukiuensis

Pleuroziaceae
Pleurozia gigantea
Pleurozia giganteoides

Lejeuneaceae
Colura ari
Colura corynephora
Diplasiolejeunea brachyclada
Diplasiolejeunea rudolphiana

Bryoxiphiaceae
Bryoxiphium norvegicum
Bryoxiphium norvegicum subsp. *japonicum*

Calymperaceae
Mitthyridium fasciculatum

Orthotrichaceae
Groutiella tomentosa

Pterobryaceae
Pseudopterobryum tenuicuspis

Sematophyllaceae
Foreauella orthothecia

Amblystegiaceae
Sinocalliergon satoi

Map 1. Six preservation regions for rare and endangered bryophytes in China

地理底图

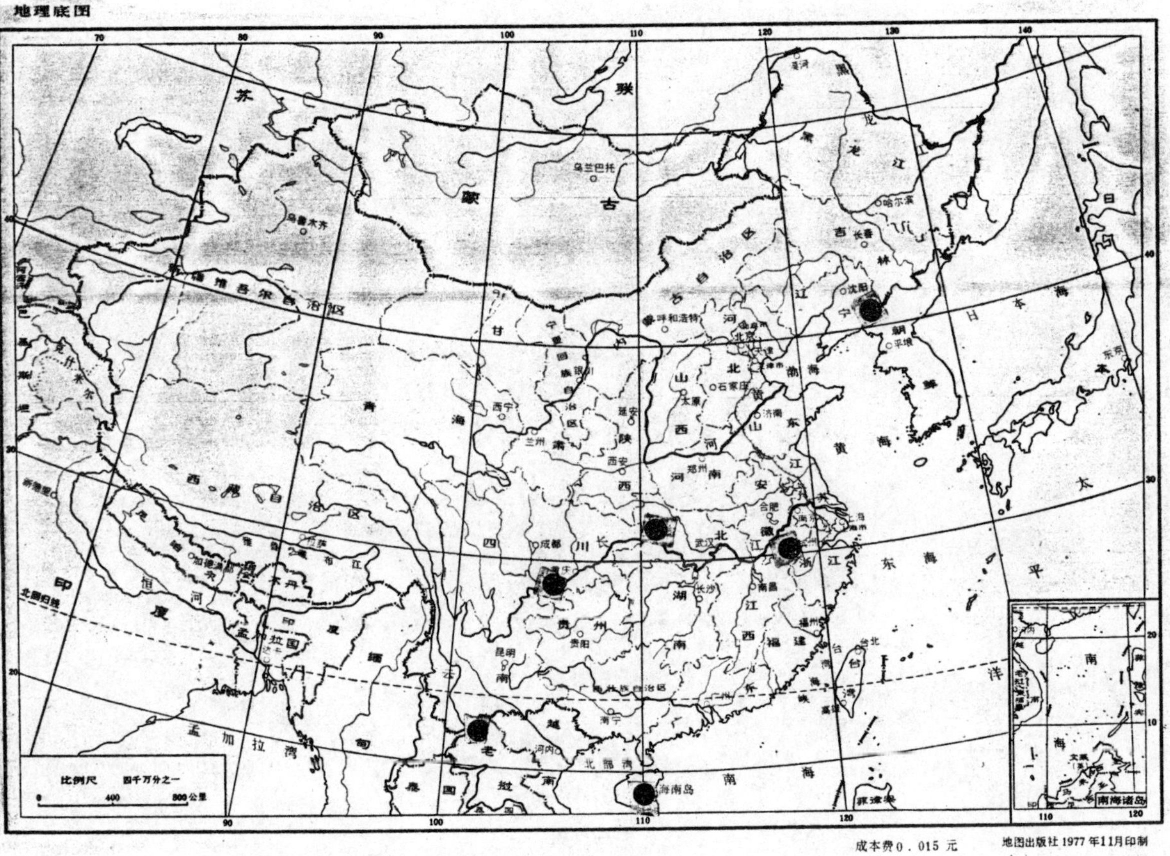

成本费0．015 元　　地图出版社 1977 年11月印制

中国苔藓植物的染色体观察Ⅲ

杜桂森
(首都师范大学生物系，北京 100037)
周云龙
(北京师范大学生物系，北京 100875)
吴鹏程 汪楣芝 贾渝
(中国科学院植物研究所 北京 100093)

CHROMOSOME OBSERVATION OF CHINESE BRYOPHYTES Ⅲ

Du Gui-sen
(Department of Biology, Capital Normal University, Beijing 100037, China)
Zhou Yun-long
(Department of Biology, Beijing Normal University, Beijing 100875, China)
Wu Pan-cheng Wang Mei-zhi Jia Yu
(Institute of Botany, Academia Sinica, Beijing 100093, China)

Abstract: Chromosome numbers are reported for 8 bryophyte species from China: *Trichocolea tomentella* (n=9); *Lepidozia reptans* (n=8+m); *Acrobolbus ciliatus* (n=9); *Gymnostomum aeruginosum* (n=13); *Neodolichomitra yunnanensis* (n=5); *Eurynchium arbuscula* (n=11); *Hypnum plumaeforme* (n=10),and *Rhytidiadelphus squarrosus* (n=9+m).
Key Words: Bryophytes; Mitosis; Chromosome number

摘要：本文报道了进行苔藓植物细胞学研究的方法和8种中国产的苔藓配子体生长点细胞有丝分裂中期的染色体数。(1)绒苔*Trichocolea tomentella* (n=9); (2)指叶苔*Lepidozia reptans* (n=9); (3)钝角顶苞苔*Acrobolbus ciliatus* (n=9); (4)铜绿净口藓*Gymnostomun aeruginosum* (n=13); (5)新船叶藓*Neodolichomitra yunnanensis* (*n*=5); (6)树状美喙藓*Eurynchium arbuscula* (*n*=11); (7)大灰藓*Hyprum plumaeforme* (*n*=10); (8)反叶拟垂枝藓*Rhytidiadelphus squarrosus* (*n*=9+m)。
关键词：苔藓植物；有丝分裂；染色体数

Materials and Methods

The material was all collected from the Wild Plants Reserve Center in Dujiangyan City, Sichuan Province, SW China except for *Gymnostomum aeruginosum* from Xishuangbanna, Yunnan Province. The plants were cultured in the medium recommended by Takio et al. (1985) in 2500-3000 lux, 14 hour light/day, and at temperature 16 ± 1°C. Observations of chromosomes were made on actively growing meristematic tissue of gametophytes as it appeared in the cultured specimens. The material was pretreated with

0.05% colchicine, and then fixed in Carnoy's fluid for 2-4 hours before it was kept in 70% ethanol at 4°C.

For observations of chromosomes, the material was transferred from the 70% ethanol to 50% ethanol, and then to distilled water. The stem tips were stripped off from the plants and macerated with the mixture of 2% pectinase and cellulase. The tips were stained with modified carbol fuchsin and then squashed. For permanent preparation, the cover slips were carefully removed after the slides were frozen for at least 2 hours. The preparations were finally remounted with gum and dried at room temperature.

Results and Discussion

1.*Trichocolea tomentella* (Ehrh) Dum.

The mitotic metaphase chromosomes in the stem tips of gametophytes were counted to be n=9 (Plate I:1). This number is in agreement with the previous reports based on the studies of populations from Japan, Poland and China's Taiwan (Tatuno, 1935,1941; Segawa, 1965; Tanaka and Inoue, 1982;Inoue et al., 1985; Kuta and Ochyra,1988). However,Newton (1971) reported n=8+m for a German population and Fritsch (1987) reported n=18 for a Caucasian population.

2.*Lepidozia reptans* (L.) Dum.

The mitotic metaphase chromosomes in the stem tips of gametophytes were counted to be n=8+m (Plate 1:2).Fritsch (1975) reported n=9 and n=10 respectively for two German populations, and Ochyra and Kuta (1985) reported n=9 for a Polish population.

3.*Acrobolbus ciliatus* (Mitt.) Schiffn.

The mitotic metaphase chromosomes in the stem tips of gametophytes were counted to be n=9 (Plate I:3). This is the first report of the chromosome count for this species.

4.*Gymnostomum aeruginosum* Sm.

The mitotic metaphase chromosomes in the stem tips of gametophytes were counted to be n=13 (Plate I:4). This number agrees with all the previous reports for American, British, Indian and Polish populations (Ireland, 1965;Smith and Newton, 1968; Kumar and Garg, 1974; Kumar and Verma, 1975; Przywara and Kuta, 1983; Newton, 1984).

5.*Neodolichomitra yunnanensis* (Besch.) T. Kop.

The mitotic metaphase chromosomes in the stem tips of gametophytes were counted to be n=5 (Plate I:5). This species is endemic to East Asia and this is the first report of its chromosome count.

6.*Eurhynchium arbuscula* Broth.

The mitotic metaphase chromosomes in the stem tips of gametophytes were counted to be n=11 (Plate I:6). This is the first report of chromosome number for this species. In this genus, *E. eustegium* (Besch.) Dix. and *E. riparioides* (Hedw.) Richards, were reported to have the same chromosome number (Yano, 1957; Du et al., 1994).

7.*Hypnum plumaeforme* Wils.

The mitotic metaphase chromosomes in the stem tips of gametophytes were counted to be n=10 (Plate I:7). This result agrees with the previous reports by Yano (1952, 1955) and that by Sannomiya (1957) for Japanese populations. However, Yano (1950) reported n=9 also for a Japanese population.

8.*Rhytidiadelphus squarrosus* (Hedw.) Warnst.

The mitotic metaphase chromosomes in the stem tips of gametophytes were counted to be n=9+m (Plate I:8). Wigh and Strandhede (1971),Inoue (1971) and (1982) reported n=10 in this species, and Newton (1973) reported n=6 for a British population.

表1. 8种苔藓植物的染色体数

Tab.1. Chromosome numbers of 8 Chinese Bryophytes

种 名 Species	科 名 Family	染色体数 Chromosome number	凭证标本 Voucher	图号 Fig.
绒苔 *Trichocolea tomentella*	绒苔科 Trichocoleaceae	n=9	Wang 51130	1:1
指叶苔 *Lepidozia reptans*	指叶苔科 Lipidoziaceae	n=9	Wang 51143	1:2
钝角顶苞苔 *Acrobolbus ciliatus*	顶苞苔科 Acrobolbaceae	n=9	Wang 51141	1:3
铜绿净口藓 *Gymnostomum aeruginosum*	丛藓科 Pottiaceae	n=13	Jia J1214	1:4
新船叶藓 *Neodolichomitra yunnanensis*	船叶藓科 Lembophyllaceae	n=5	Wang 51144	1:5
树状美喙藓 *Eurhynchium arbuscula*	青藓科 Brachytheciaceae	n=11	Wang 51138	1:6
大灰藓 *Hypum plumaeforme*	灰藓科 Hypnaceae	n=10	Wang 51116	1:7
反叶拟垂枝藓 *Rhytidiadelphus squarrosus*	垂枝藓科 Rhytidiaceae	n=10	Wang 51134	1:8

参考文献

[1] 李懋学、张方. 1991. 植物染色体研究技术，东北林业大学出版社，第一版.

[2] 周云龙、王长锐、沙伟. 1989，关于苔藓植物染色体研究方法的探讨. 武汉植物研究，7(2):115－118.

[3] 吴鹏程、郑培中. 1989. 有关中国苔藓植物染色体数目资料，考察与研究增辑，83－121.

[4] 吴鹏程，1992. 苔藓植物研究的新进展. 植物分类学报，30(2):183－192.

[5] Takio, S., S. Takani & S.Hino. 1985. Nitrate reductase from thalli of *Marchantia polymorpha*. Journ Hattori Bot. Lab. 58:131-137.

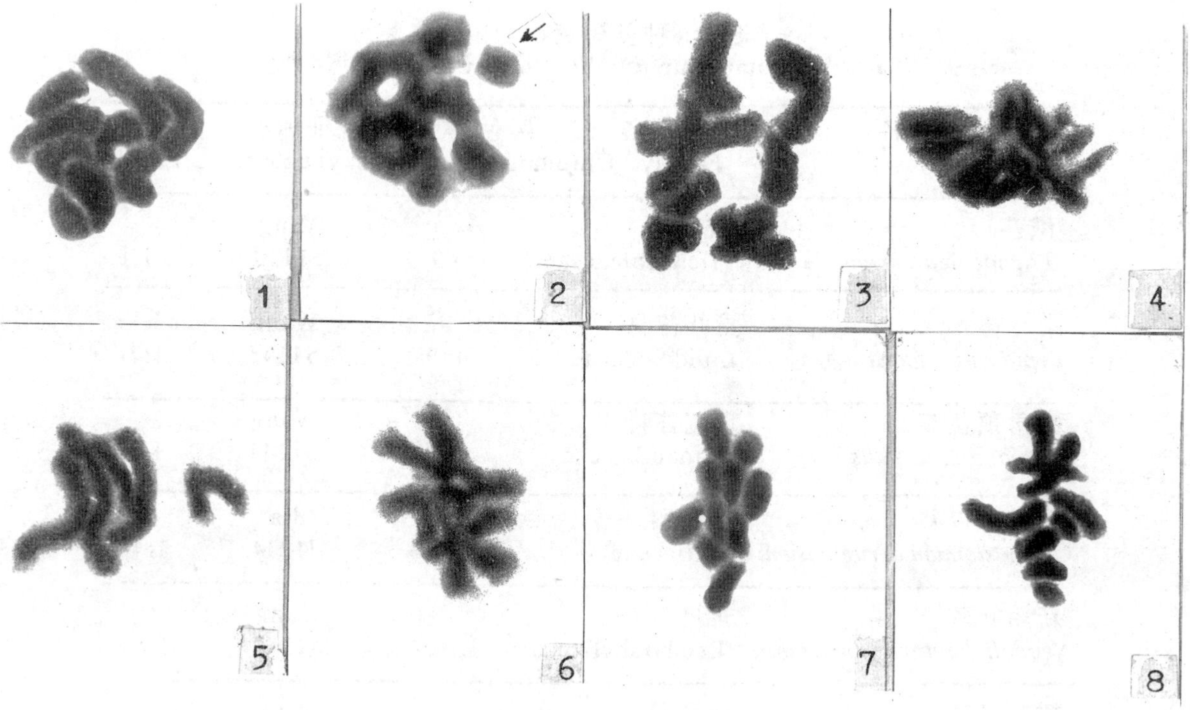

Plate 1.

Chromosome numbers of 8 Chinese bryophytes.

中国苔藓植物染色体观察 Ⅳ

周云龙 马丽霞* 何强 孙永见 江红芳
(北京师范大学生物系 北京 100875)
吴鹏程 汪楣芝
(中国科学院植物研究所 北京 100093)
杜桂森
(首都师范大学生物系 北京 100037)

CHROMOSOME OBSERVATION OF CHINESE BRYOPHYTES Ⅳ

Zhou Yun-long Ma li-xia He Qiang Sun Yong-jian Jiang Hong-fang
(Department of Biology, Beijing Normal University, Beijing 100875, China)
Wu Pan-cheng Wang Mei-zhi
(Institute of Botany, Academia Sinica, Beijing 100093, China)
Du Gui-sen
(Department of Biology, Capital Normal University, Beijing, 100037, China)

Abstract: The chromosome numbers of eight Chinese bryophytes are reported: (1) *Dumortiera hirsuta* (SW.) Reinw., Bl. & Nee, n=27; (2) *Marchantia tosana* Steph., n=9; (3) *Grimmia pilifera* P.Beauv., n=9; (4) *Atrichum undulatum* (Hedw.) P.Beauv., n=7; (5) *Hypnum* sp., n=10; (6) *Bryum argenteum* Hedw., n=10; (7) *Physcomitrium sphaericum* (Ludw.) Fuernr., n=48; and (8) *Pterygoneurum ovatum*, n=7.

Key words: Bryophytes; Chromosome number; Mitosis; Meiosis

摘要: 本文报道了中国产的8种苔藓植物的染色体数. 1. 毛地钱 *Dumortiera hirsuta* (Sw.) Reinw., Bl. et Nee. n=27; 2.东亚地钱*Marchantia tosana* Steph. n=9; 3.毛尖紫萼藓*Grimmia pilifera* P.Beauv. n=9; 4.仙鹤藓 *Atrichum undulatum* (Hedw.) P.Beauv. n=7;.灰藓属之一种*Hypnum* sp. n=10; 6.真藓*Bryum argenteum* Hedw. n=10; 7.立碗藓 *Physcomitrium sphaericum* (Ludw.) Fuernr. n=48; 8.卵叶盐土藓*Pterygoneurum ovatum* (Hedw.) Dix. n=7.

关键词: 苔藓植物; 染色体数; 有丝分裂; 减数分裂

一、材料和方法

8种苔藓植物于1994—1995年分别采于中国昆明、江西、山东、福建、韶关和新疆等地. 研究方法主要参见周云龙等[1][2]的报道. 其中6种材料为配子体(n), 观察其茎尖或生长点细胞有丝分裂中期的染色体数, 除卵叶盐土藓是将其配子体在室内进行培养待长出新的茎尖予以固定外, 其余5种材料均是在采集地点当即用卡诺固定液固定．另有2种材料是观察孢子体($2n$)的孢蒴中孢子母细胞减数分裂中期1的二倍体．染色体制片均在Olympus显微镜的油镜下进行显微摄影．

二、结果与讨论

8种苔藓植物的染色体数见表1.

表1. 中国8种苔藓植物的染色体数

Tab.1. Chromosome number of 8 Chinese Bryophytes

种 名 Species	科 名 Family	染色体数 Chromosome number	产地	凭证标本 Voucher	图号 Fig.
毛地钱 *Dumortiera hirsuta*	地钱科 Marchantiaceae	n=27	江西	Qiang H. J9401	I:1
东亚地钱 *Marchantia tosana*	地钱科 Marchantiaceae	n=9	江西	Qiang H. J9404	I:2
毛尖紫萼藓 *Grimmia pilifera*	紫萼藓科 Grimmiaceae	n=9	山东牟平	Yongjian S. SD9402	I:3
仙鹤藓 *Atrichum undulatum*	金发藓科 Polytrichaceae	n=7	福建连城	Hongfang J. F9403	I:4
灰藓属之一种 *Hypnum* sp.	灰藓科 Hypnaceae	n=10	山东牟平	Yongjian S. SD9401	I:5
真藓 *Bryum argenteum*	真藓科 Bryaceae	n=10	云南昆明	Fu S. F9400	I:6
立碗藓 *Physcomitrium sphaericum*	葫芦藓科 Funariaceae	n=48	广东韶关	Lixia M. S9500	I:7
卵叶盐土藓 *Pterygoneurum ovatum*	丛藓科 Pottiaceae	n=7	新疆	Hong W. X9400	I:8

上述8种苔藓植物中有3种的染色体数为首次报道，即卵叶盐土藓、立碗藓和毛尖紫萼藓（图I:3）．真藓（图I:6）、仙鹤藓（图I:4）、东亚地钱（图I:2）和已有的报道一致（吴鹏程、郑培中，1989）．毛地钱（图I:1）和已有的报道基本相同，Tatuno (1965)（吴鹏程、郑培中，1989；Fritsch，1982)曾研究该种的染色体数为n=9, 18, 27, 本文的报道为n=27，是n=9的三倍体．至于1种灰藓与灰藓属中多数种类相同，为n=10, 但该属中有些种为n=6，7，8，12和14（吴鹏程、郑培中，1989）．

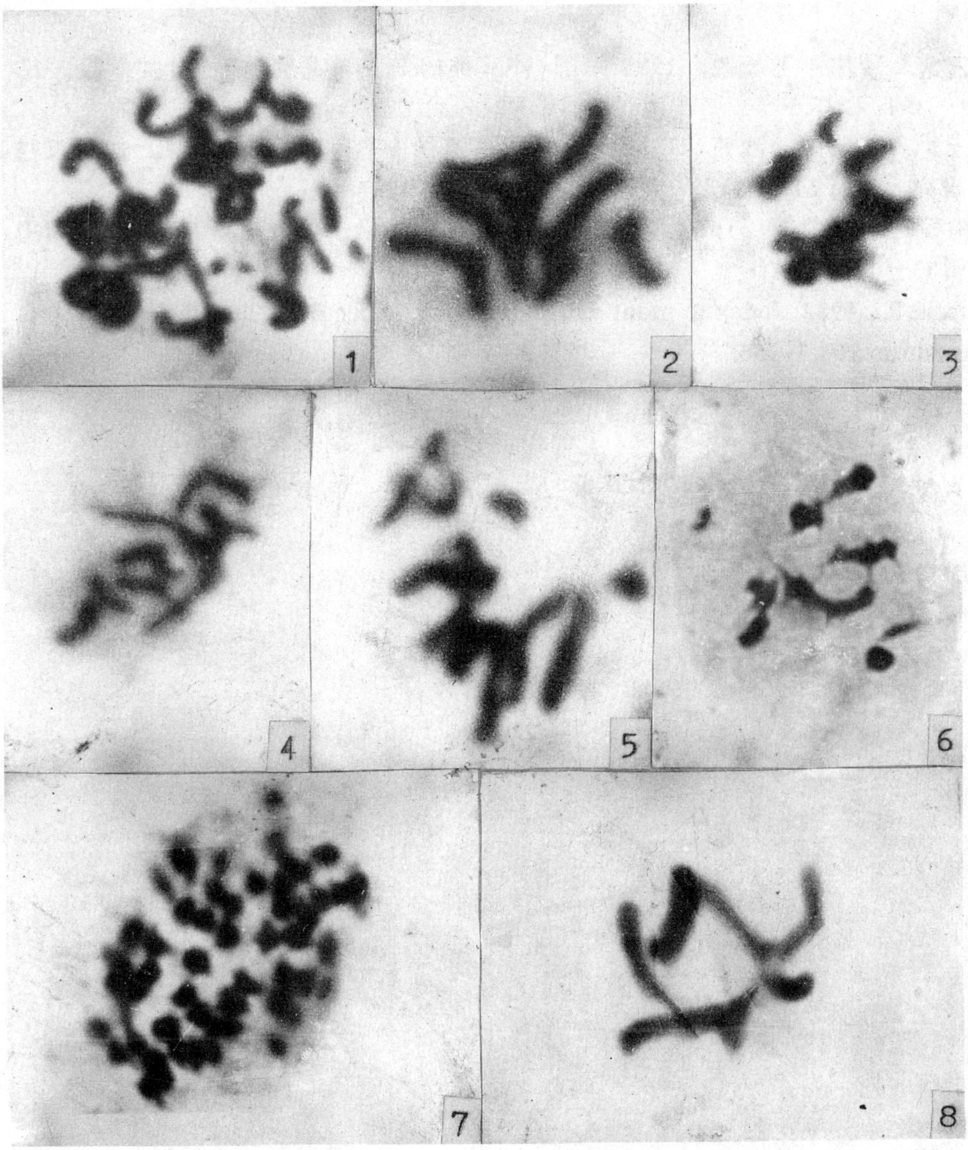

图版I: 1–8 中国苔藓植物的染色体观察IV

P1. I:1-8. Observation of chromosomes of Chinese bryophytes

1.毛地钱*Dumortiera hirsuta* (Sw.) Reinw., Bl. et Nee. n=27 (× 2200)

2.东亚地钱*Marchantia tosana* Steph. n=9 (× 3300)

3.毛尖紫萼藓*Grimmia pilifera* P.Beauv. n=9(× 2600)

4.仙鹤藓*Atrichum undulatum* (Hedw.) P.Beauv. n=7 (× 3500)

5.灰藓一种*Hypnum* sp. n=10 (× 3300)

6.真藓*Bryum argenteum* Hedw. n=10 (× 1800)

7.立碗藓*Physcomitrium sphaericum* (Ludw.) Fuernr. n=48 (× 1600)

8.卵叶盐土藓*Pterygoneurum ovatum* (Hedw.) Dix. n=7 (× 3500)

参考文献

[1] 周云龙、沙伟、王长锐，1988:中国4种藓类的细胞学观察. 植物分类学报，26(5): 378－381.

[2] 周云龙、王长锐、沙伟，1989: 关于苔藓植物染色体研究方法的探讨. 武汉植物学研究，7(2):115－118.

[3] 吴鹏程、郑培中等，1989: 有关中国苔藓植物的染色体数目资料. 考察与研究增辑83－121.

[4] Fritsch, R., 1982: Index to plant chromosome number Bryophytes. Regnum Vegetabile. 108:1-256.

蛇苔属*Conocephalum*植物的两种细胞染色体核型的研究

杨晓杰　沙伟
(齐齐哈尔师范学院生物系, 齐齐哈尔 161006)
汪楣芝
(中国科学院植物研究所,北京 100093)

KARYOLOGICAL OBSERVATIONS ON THE GENUS *CONOCEPHALUM* WIGG.FROM CHINA

Yang Xiao-jie and Sha Wei
(*Department of Biology, Qiqihare Teacher's College, Qiqihare,161006, China*)
Wang Mei-zhi
(*Institute of Botany, Academia Sinica, Beijing, 100093, China*)

Abstract: Chromosome numbers and karyotypes of the only two species in the small genus *Conocephalum* Wig. are reported in this paper. *C.conicum* had the karyotype of K(*n*)=8=5m+3sm or K(*n*)=8=5V+3(v), and *C. japonicum* had the karyotype of K(*n*)=8=5m+3sm or K(*n*)=8=5V+2J+1(v).
Key words: *Conocephalum* Wig.; Chromosome; Karyotype

摘要: 本文作者对中国浙江西天目山的两种蛇苔植物进行了研究，其结果染色体数目均为*n*=8. 按照李懋学（1991）的核型公式，它们的排列为K（*n*）=8=5m+3sm，参考Inoue（1983）的标准，蛇苔*C.conicum*为K(*n*)=8=5V+3(v)，而小蛇苔*C.japonicum*为K(*n*)=8=5V+2J+1(v). 两种蛇苔的核型在中国为首次报道.
关键词: 蛇苔属; 细胞染色体; 核型

Conocephalum Wigg.is a small liverwort genus of only two species, *C.conicum* and *C.japonicum*.This genus is widely distributed in the northern hemisphere, growing usually on moist soils or rock under forest.

Numerous karyological studies have been made on this genus (Newton,1990;Anand et al., 1989; Inoue et al., 1985; Kuta et al., 1984; Oosaki,1980; Tatuno, 1938a, 1938b, 1941, 1957),karyological data on this genus from China, however, are absent in standard references. In this paper, chromosome numbers and karyotypes of the two species in this genus from China are reported.

Materials and Methods

Information of the materials studied here is listed in Table 1. The Plants were collected on the rocks at an elevation of 1000m of Tianmu Mt., Zhejiang Province. The voucher specimens were deposited in the Herbarium of Qiqihare Teacher's College (QTC) and the Herbarium of Institute of Botany, Academia Sinica (PE).

Table 1. Chromosome numbers for two *Conocephalum* species

Species	Locality	Voucher	Chromosome number
C.conicum	Mt. Tianmu, Zhejiang Province	Mei-Zhi Wang, Wei Sha 95083	8
C.japonicum	Mt. Tianmu, Zhejiang Province	Mei-Zhi Wang, Wei Sha 95059	8

For chromosome observations,apical growth tips of the gametophytes were pretreated with saturated P-Dichlorobenzene solution for 3 hours, they were then fixed in Carnoy's fluid for 1 to 2 hours. After being macerated in the 1:1 mixture of 95% ethanol: HCl for 10-15 minutes, the tips were stained with carbol fuchsin and then squashed.

The method of karyotype analysis followed Inoue (1983),and Li and Cheng (1985).

Results and Discussion

1. *Conocephalum conicum* **(L.) Dum.**

The interphase nuclei were categorized as simple chromocenter type (Plate I:1). The metaphase chromosomes were counted to be n=8 (Fig.3;Plate I:3), the largest being 3.45μm in length, the smallest 0.95μm. The total absolute length of the chromosome complement was 17.36μm. The relative length of the chromosomes ranges form 5.47 to 19.87,and the arm ratio ranges from 1.07 to 1.95.The chromosome complement was composed of 8 metacentric chromosomes and 3 submetacentric ones,thus the karyotype was formulated as K(n)=8=5m+3sm or K(n)=8=5V+3(v). The karyogram is shown in Plate I:5,the idiogram in Fig.1, and the parameters of the chromosomes in Table 2.

The present number n=8 is in agreement with the previous report by Anand et al. (1989) for a population from India and that by Inoue (1985) for a population from Japan, but they all reported the karyotype as K(n)=4sm+4st.

The other two chromosome numbers reported in this species are n=9 and n=8+m for Polish, Indian and Japanese populations (Newton,1990;Inoue,1985; Kumar and Anand, 1984; Kuta, 1984; Oosaki, 1980; Tatuno, 1938a, 1941).

2. ***Conocephalum japonicum*** **(Thunb.) Grolle**

The interphase nuclei were categorized as simple chromocenter type (Plate I:2).

Table 2. The Parameters of Chromosomes of Two *Conocephalum* Species

Species	No.	Absolute chromosome length(μm) (short arm+long arm)	Relative length	Arm ratio	Form%	Typ e	
	1	1.67 + 1.78	19.87	1.07	48.41	∇	m
	2	1.10 + 2.15	18.72	1.95	33.85	∇	sm
	3	1.23 + 1.49	15.67	1.21	45.22	∇	m
C.	4	0.90 + 1.75	15.27	1.94	33.96	∇	sm
conicum	5	0.72 + 1.28	11.52	1.78	36.00	∇	sm
	6	0.55 + 0.81	7.83	1.47	40.44	(V)	m
	7	0.41 + 0.57	5.65	1.39	41.84	(V)	m
	8	0.83 + 0.57	5.47	1.50	40.00	(V)	m
	1	0.68 + 1.03	18.67	1.51	39.77	∇	m
	2	0.41 + 1.05	15.94	2.56	28.08	J	sm
	3	0.61 + 0.72	14.52	1.20	45.86	∇	m
C.	4	0.52 + 0.71	13.43	1.37	42.28	∇	m
japonicum	5	0.50 + 0.60	12.00	1.20	45.45	∇	m
	6	0.41 + 0.49	9.83	1.20	45.56	∇	m
	7	0.28 + 0.59	9.50	2.11	32.18	J	sm
	8	0.20 + 0.36	6 11	1.80	35.71	(V)	sm

The metaphase chromosomes were counted to be n=8 (Fig.4;Plate I:4), the largest being 1.71μm in length, the smallest 0.56μm. The total absolute length of the chromosome complement was 9.16μm. The relative length of the chromosome ranges from 6.11 to 18.67,and the arm ratio ranges from 1.20 to 2.56.Among the 8 chromosomes of the chromosome complement, 5 were metacentric and the remaining 3 submetacentric, thus the karyotype could be formulated as K(n)+8=5m+3sm or K(n)=8=5V+2J+1(v)> The karyogram is shown in Plate I:6, the idiogram in Fig. 2, and the parameters of the chromosomes in Table 2.

The number n=8 is a new chromosome count for this species. In Japanese populations of this species, n=9 and n=8+m were reported (Tatuno, 1957, 1938b, 1941).

Compared with those of *C. conicum*, the chromosomes of this species are obviously smaller.The difference of the total absolute length of the chromosome complements between the two species reaches 8.20μm.

The idiograms of two *Conocephalum* species
两种蛇苔的染色体

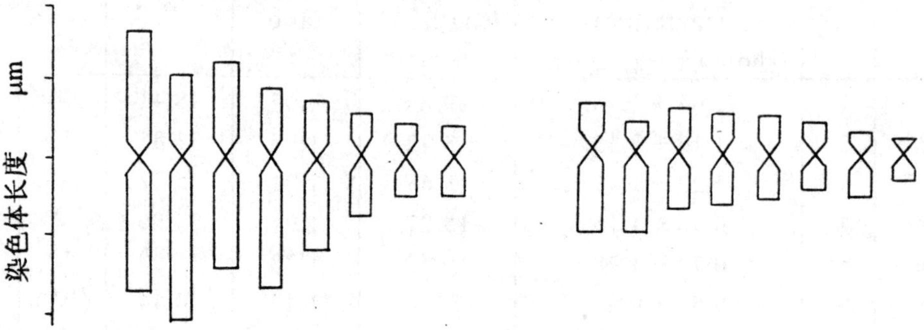

Fig.1 Fig.2

Fig.1 *Conocephalum conicum* (L.) Dum. 蛇苔
Fig.2 *C. japonicum* (Thunb.) Grolle 小蛇苔

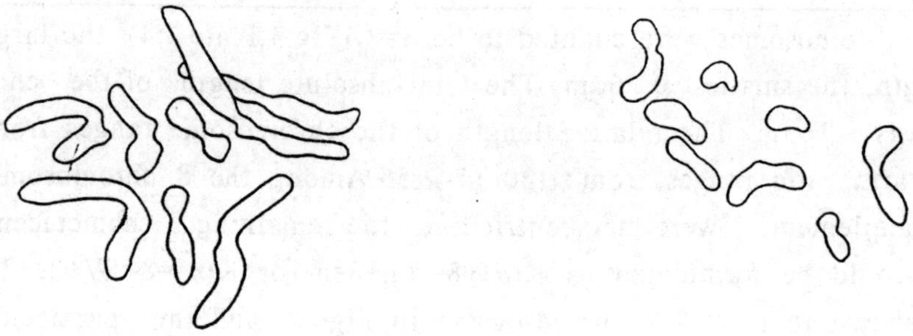

Fig.3 Fig.4

Fig.3 *Conocephalum conicum* (L.) Dum. 蛇苔
Fig.4 *C. japonicum* (Thunb.) Grolle 小蛇苔

Explanatory diagrams of Plate I: 1–6

图版说明 I: 1–6

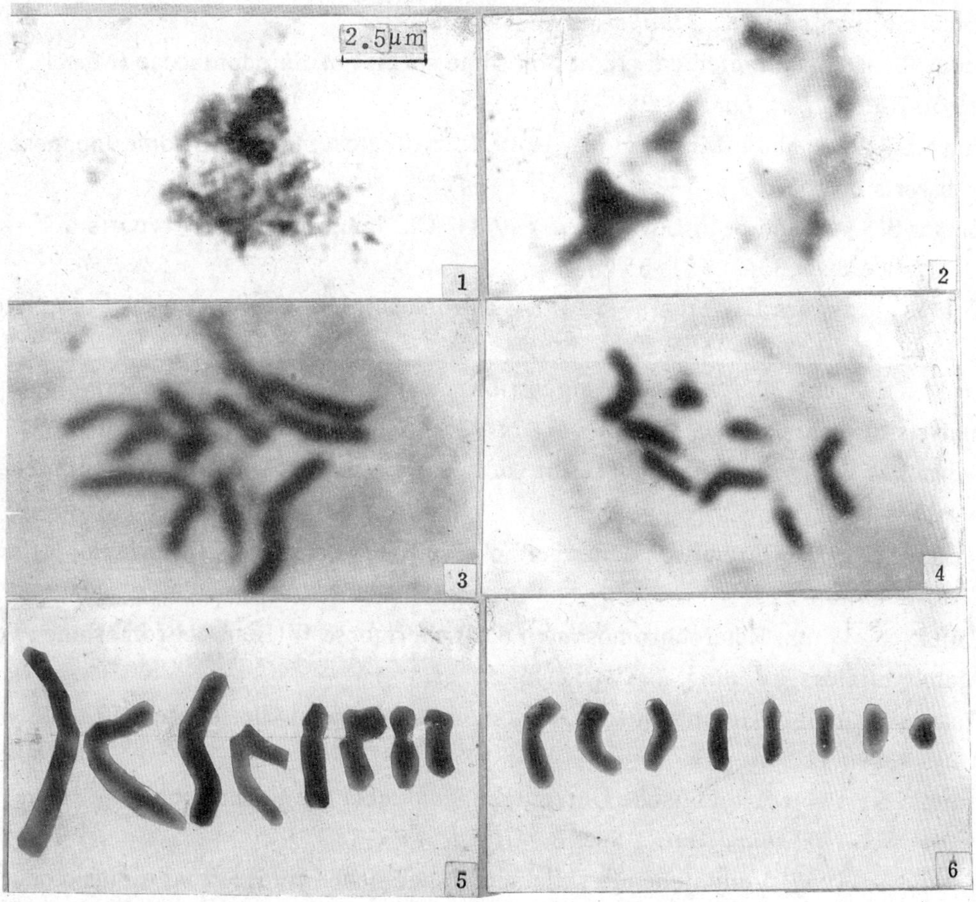

Fig.1.3.5 *Conocephalum conicum* (L.) Dum. 蛇苔

Fig.2.4.6 *C. japonicum* (Thunb.) Grolle 小蛇苔

Fig.1-2 resting stage; 3-4 mitotio metaphaphase; 5-6 Katyotype

REFERENCES

[1] Anand S., Kumar S.S., Kaur B., 1989. Cytological observations on some West Himalaya liverworts Ⅱ. *Lindbergia* 15:99–102.

[2] Inoue S., 1983. Karyological studies on some species of Entodontaceae (Musci). *Journ. Hattori Bot. Lab.* 54:299–305.

[3] Inoue S., Watanabe I.,Mizutani M., 1985. Karyological studies on some Japanese liverworts. *Hikobia* 9:235–241.

[4] Kumar S.S., Anand S. in Love A (ed.). 1984. Chromosome number reports LXXXIII. *Taxon* 33(2):351–354.

[5] Kuta E., Przywara L., Ochyra R., 1984. Chromosome studies on polish bryophytes I. *Bryol. Beitr.* 3:28–45.

[6] Li M.X., Cheng R. Y., 1985. A suggestion on the standardization of karyotype analysis in plants. *Journ. Wuhan Bot. Res.* 3(4):297–302.

[7] Newton M.E., 1990. Genetic structure of hepatic species. *Bot. Journ. Linn. Soc.* 104:215–229.

[8] Oosaki M.,1980. Chromosome numbers of five liverwort species from Japan. *Misc. Bryol. Lich.* 8(7):136.

[9] Tatuno S., 1938a. Heterochromosomen der Lebermoose Ⅳ.Heterochromosomen von Marchantiales. *Bot. Mag. Tokyo* 52:45–51.

[10] Tatuno S., 1938b. Geschlechtschromosomen bei einigen Lebermoosen. Ⅷ. Bot. *Mag. Tokyo* 52:374–379.

[11] Tatuno S., 1941. Zytologische Untersuchungen ueber die Lebermoose von Japan. *Journ. Sci. Hiroshima Univ.*, Ser. B., Div. 2, 4:73–187.

[12] Tatuno S., 1957. Vergleichende Untersuchungen ueber die Heterochromasie bei Ordnung Marchantiales Ⅱ. *Journ. Sci. Hiroshima Univ.*,Ser. B. Div. 2,8:81–116.

培养基pH值对苔藓孢子萌发及原丝体生长的影响*

张朝晖 刘宁 钟本固 陈蓉蓉 王智慧
(贵州师范大学生物科学技术系 贵阳550001)

EFFECT OF pH VALUES OF GROWTH MEDIUM ON SPORE GERMINATION AND PROTONEMAL GROWTH OF THREE SPECIES OF BRYOPHYTES COLLECTED FROM KARST AREA OF GUIZHOU PROVINCE, SW CHINA

Zhang Zhao-hui, Liu Ning, Zhong Ben-gu, Chen Rong-rong and Wang Zhi-hui
(Department of Biological Science and Biotechnology, Guizhou Normal University, Guiyang 550001, China)

Abstract: Spore germination and protonema growth of three species of bryophytes in response to different pH values of growth medium were conducted. Specimen of *Plagiochasma sessillicephalum* Horik.was collected from Huangguoshu karst area and specimens of *Timmiella anomala* (B.S.G.) Limpr.and *Bartramia ithyphylla* Brid. came from forested karst area in Yuntai Hill, Guizhou province. Data of spore germination were taken from 3-days old cultures and the protonemal growth data from 10-days old cultures.

Under cultivation,spores of *Timmiella anomala* showed germination at pH 6-10, spores of *Bartramia ithyphylla* germinated at pH 5-10, and that of *Plagiochasma sessillicephalum* at pH 4-10. Maximal percentage of spore germination in *Timmiella anomala* and *Bartramia ithyphylla* was observed at pH 8, and that of *Plagiochasma sessillicephalum* at pH 6. Maximal protonemal growth in *Timmiella anomala* was observed at pH 8,*Bartramia ithyphylla* at pH 9,and *Plagiochasma sessillicephalum* at pH 6.

Key Words: pH; Spore; Bryophytes; *Timmiella; Bartramia; Plaqiochasma.*

摘要: 本文作者在室内采用液体培养基研究了pH浓度对紫背苔*Plagiochasma sessillicephalum* Horik.、反纽藓*Timmiella anomala* (B.S.G.) Limpr.和直叶珠藓 *Bartramia ithyphylla* Brid.的影响. 实验材料采自贵州黄果树喀斯特区域和云台山喀斯特林区. 培养基pH浓度梯度为3、4、5、6、7、8、9和10. 研究结果表明, 短柄紫背苔孢子萌发范围为pH=4-10, 反纽藓pH=6-10, 直叶珠藓pH=5-10; 短柄紫背苔孢子萌发率最适点pH=6, 反纽藓和直叶珠藓最适点pH=8; 紫背苔原丝体最适生长点pH=6, 反纽藓pH=8, 直叶珠藓pH=9.

关键词: pH; 苔藓植物; 孢子

培养基pH浓度对苔藓植物生长的影响, 曾被国外许多学者研究过(野口彰, 1959; K.S. Vishvakarma & A. Kaul, 1988), 但国内有关这方面的实验文献少见报道(吴鹏程,

1993;Gao & Cao,1989), 为了探讨喀斯特地区苔藓植物的适应机制，有必要对此问题进行实验探讨.

一、研究材料及方法

(一)植物材料: ①短柄紫背苔*Plagiochasmase ssillicephalum* Horik. No.3286采自贵州黄果树喀斯特区域，基质钙质土，pH=6.5;②反纽藓*Timmiella anomala* (B.S.G.) Limpr. No.1733采自贵州施秉云台山喀斯特区域，基质为常绿落叶阔叶林下白云岩或钙质土，pH=6.5-7. ③直叶珠藓*Bartramia ithyphylla* Brid. No1791采自贵州施秉云台山喀斯特区域，基质为常绿落叶林下钙质土，pH=6.5.

上述三种苔藓植物均有含有大量成熟孢子的孢蒴，标本存于贵州师大生物科学技术系植物标本室.

(二)培养基的配制: 按一定比例配制液体培养基(周云龙，1988). 采用H_3PO_4调节pH,设置3、4、5、6、7、8、9和10等8个浓度梯度.

本实验采用处理过的瓦片作为液体培养基中孢子培养的载体. 瓦片具有易取材、易观察等优点,并且自身具多毛细孔道, 吸水力强, 较好地解决了培养中的水、气矛盾. 处理方法如下: 将面积3 × 4cm^2的瓦片洗净，酸处理3小时；把酸冲洗干净，蒸馏水中静置24小时；冲洗后测定pH=7, 85℃恒温24小时烘干.

在无菌条件下，将不同pH浓度液体培养液倒入直径12cm的玻璃培养皿中，每个培养皿中轻放一块瓦片，液体接近但不能超过瓦片表面. 设置8个浓度梯度,三组重复对比.

(三)接种: 取已消毒的三种苔藓孢子制成孢子悬液，在超净工作台内接种在瓦片上，盖好培养皿并用透明胶布粘封. 接种时间是1994年4月28日.

(四)培养: 置培养皿于实验室工作台上, 静置培养. 白天自然光照, 晚上日光灯照射. 培养期间室内环境因子统计平均变化量如图一所示.

(五)观察: 第3天抽样制片观察记录孢子萌发率，第10天制片观察并测量原丝体长度.

二、实验结果

1.pH值与苔藓植物孢子萌发范围:

实验结果表明(表1): 不同的苔藓植物孢子萌发范围的pH浓度不一样. 短柄紫背苔孢子萌发范围为pH=4-10，反纽藓pH=6-10，直叶珠藓pH=5-10. 喀斯特地区生长的苔藓植物，表现了对碱性环境的适应,而在环境较酸的条件下(如短柄紫背苔pH<4, 反纽藓pH<6，直叶珠藓pH<5), 则孢子不能正常萌发.

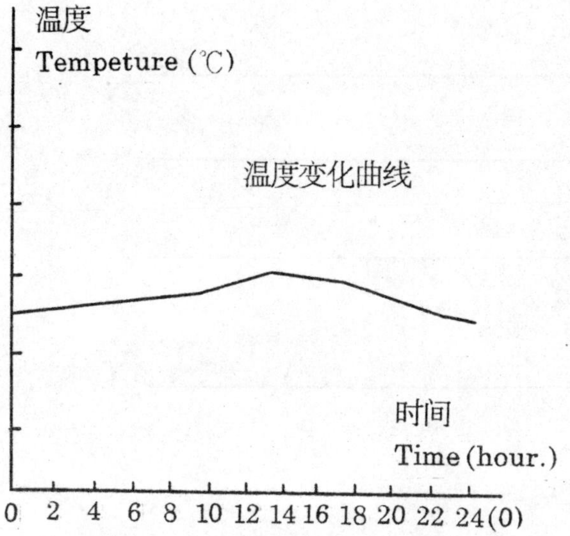

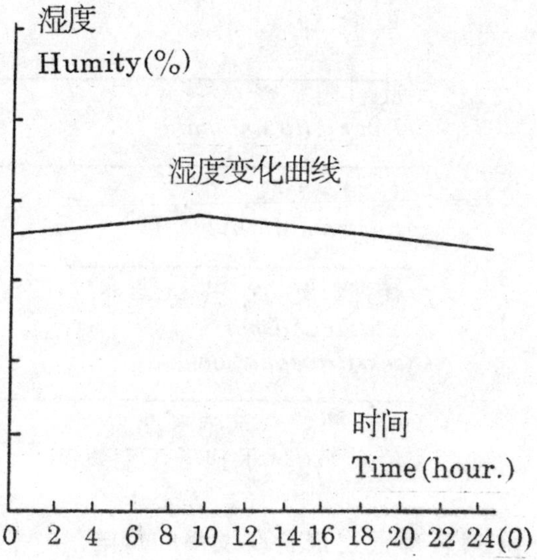

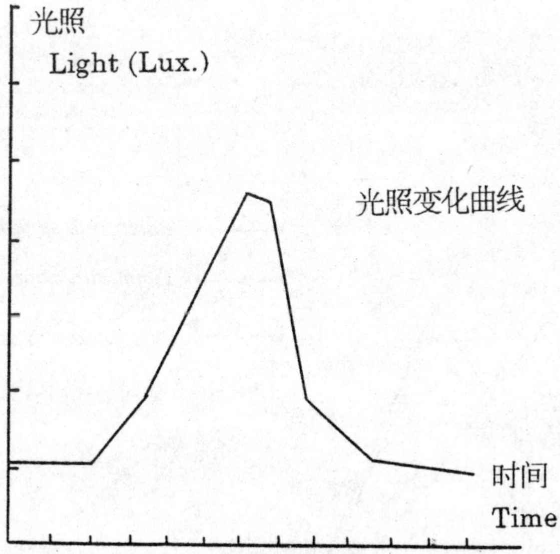

图1、苔藓孢子萌发的实验室环境因子

Fig.1:Enviromental factors for spore germination of bryophytes in laboratory

表1、pH值对三种苔藓植物孢子萌发的影响

Table 1. Effect of pH on spore germination of three bryophytes

pH 苔藓种类	3	4	5	6	7	8	9	10
反纽藓 *Timmiella anomala*	×	×	×	✓	✓	✓	✓	✓
直叶珠藓 *Bartramia ithyphylla*	×	×	✓	✓	✓	✓	✓	✓
短柄紫背苔 *Plagiochasma sessillicephalum*	×	✓	✓	✓	✓	✓	✓	✓

注：“✓”示孢子萌发；
“×”示孢子不萌发.

2.pH对苔藓植物孢子萌发率的影响：

根据不同pH梯度下苔藓植物孢子萌发量的统计平均数，绘出三种苔藓植物孢子萌发率曲线(图2)．从曲线中可以看到，不同苔藓孢子在萌发时对pH反应不一样．短柄紫背苔pH=6时孢子萌发率最高，而反纽藓和直叶珠藓则在pH=8时孢子萌发率最高．

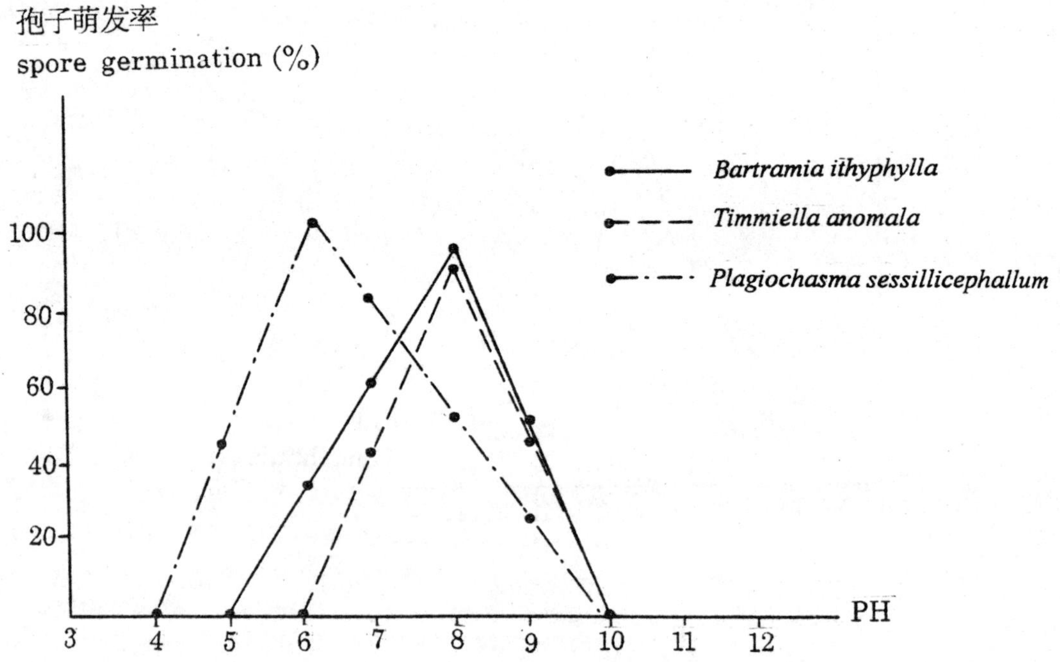

图2、pH值对三种苔藓植物孢子萌发率的影响

Fig.2:Percentage of spore germination of three bryophytes at different pH

3.pH值对苔藓植物原丝体生长的影响:

从原丝体生长状况来看(表2):反纽藓原丝体生长适宜范围pH=7、8、9,最适点pH=8;直叶珠藓pH=8、9、10, 最适点pH=9; 短柄紫背苔pH=6、7、8, 而pH=6时为最适点.

上述三种苔藓原丝体的生长，均表现了它们对环境偏碱性或中性的适应．由于大多数裸露的喀斯特生境都具有微碱性或中性的特征(贵州环境科学会主编，1988; 朱守谦等, 1993)，喀斯特地区苔藓植物这一特点，对其生存具有极重要的生理适应意义.

表2、pH值对苔藓植物原丝体生长的影响

Table 2: Effect of pH on protonema growth of three bryophytes

苔藓植物名称	原 丝 体	pH 3	4	5	6	7	8	9	10
直叶珠藓	原丝体总长度(μm)	/	/	270.6	295.2	295.2	319.8	639.6	369
Bartramia ithyphylla	原丝体细胞长度(μm)	/	/	24.6	36.9	36.9	49.2	135.3	49.2
	原丝体细胞宽度(μm)	/	/	4.92	4.92	7.36	12.3	9.84	7.36
反 纽 藓	原丝体总长度(μm)	/	/	/	98.4	147.6	295.2	221.4	86.1
Timmiella anomala	原丝体细胞长度(μm)	/	/	/	12.3	17.22	49.2	36.9	24.6
	原丝体细胞宽度(μm)	/	/	/	2.46	4.92	9.84	7.38	4.92
紫 背 苔 短柄	原丝体总长度(μm)	/	49.2	59.04	78.72	73.8	68.88	66.42	54.12
Plagiochasma sessilli-cephalum	原丝体细胞长度(μm)	/	19.68	19.68	24.6	22.14	19.68	9.84	7.38
	原丝体细胞宽度(μm)	/	7.38	7.38	17.22	7.38	4.92	2.46	2.46

三、讨论

野口彰等(1959)在对日本石灰岩地区苔藓植物进行培养实验后指出，苔藓植物的生长与氢离子浓度有明显的相关性．国外其他研究者的工作，如Woodifin(1976) ,K.S. Vishvakarma 和 A. Kaul(1988)等，支持这一见解．作者的初步实验，也证实了同样的观点．

从反纽藓、直叶珠藓和短柄紫背苔孢子的萌发及原丝体的生长来看，它们是适宜于生长在中性或微碱性生境中的植物．在野外工作中，我们常观察到三种苔藓植物配子体在贵州的主要生长基质有钙质土、石灰岩、白云岩和钙华，测得基质pH=6–7之间，中性略偏微酸，似乎与实验结果相矛盾．我们推测这种现象与苔藓植物生命代谢过程中产生有机酸改变了小生境有关．据有的地质、生物研究者报道(贵州环境科学会主编，1988; 朱守谦等，1993),在尚未有高等植物侵占的裸露碳酸岩地貌上，pH值常为7–8之间，呈中生或微碱性．

喀斯特地区苔藓植物孢子萌发及原丝体生长适宜于中性或偏微碱性环境具有重要的生理和生态意义，这使苔藓植物在生境略偏碱性或中性的喀斯特地貌上充当先锋植物，改变了裸露的碳酸盐岩石上缺乏高等植物的自然景观．

参考文献

[1] 吴鹏程，1993: 中国苔藓植物研究的回顾与展望, Chenia, Vol.I:1–6.

[2] 周云龙，1988: 孢子植物实验及实习, 北京师范大学出版社，第8页，北京.

[3] 贵州环境科学学会主编，1988: 贵州喀斯特环境研究, 贵州人民出版社.

[4] 朱守谦主编，1993: 喀斯特森林生态研究(I)，贵州科技出版社.

[5] 野口彰，1959: 日本石灰岩地带の藓类, 广岛大学生物学会志，第10卷26期.

[6] Gao,C.& T. Cao,1989: Research activities in Chinese bryology throughout 1980's. *Acta Bryolichenol, Asiatica* (1 & 2):21–30.

[7] Vishvakarma, K.S. & A. Kaul, 1988: Culture studies on *Plagiochasma appendiculatum* Lehm. et Lindenb. and *Reboulia hemisphaerica* (L.) Raddi populations of Pachmarhi (central India) in relation to PH on a comparative basis. *Cryptogamie*, 9(2): 129–135.

福建万木林保护区叶附生苔类的研究*

李登科
（上海自然博物馆，　上海 200002）

A STUDY ON EPIPHYLLOUS LIVERWORTS OF WANMULIN NATURAL RESERVE IN FUJIAN PROVINCE，E CHINA

Li Deng-Ke
（*Shanghai Museum of Natural History, Shanghai 200002, China*）

Abstract: Wanmulin Nature Reserve is situated in the southern flank of Wuyi Mountain Range in northwest part of Fujian Province, about 27°03′N and 118°09′E. The study area has a monsoonal climate with subtropical evergreen forest cover. There were no record of epiphyllous liverworts from Wanmulin Natural Reserve prior to the present study. In 1986, the author collected 23 specimens of epiphyllous liverworts. The collections include 3 families, 8 genera and 14 species representing new records for the area. Among these, *Pedinolejeunea himalayanensis* var. *fukiensis* is a newly described variety and *P.formosana* is new to Fujian Province.

This paper includes the following parts: (1) an enumeration of the epiphyllous liverworts found in Wanmulin; (2) the ecology or environment of the epiphyllous liverworts; (3) phorophyte information; (4) floristic relationship between the study area and neighboring mountain regions; (5) a discussion.
Keywords: *Pedinolejeunea*; Wanmulin, Fujian; Epiphyllous liverworts

摘要：万木林自然保护区在福建省西北部，位于武夷山南侧的低丘陵带，是中国最早建立的保护区之一．区内由亚热带常绿阔叶混交林组成，林内苔藓植物较丰富．作者于1986年8月深入保护区采集苔藓植物标本500多号，其中有叶附生苔类标本23号．经鉴定，计有3科、8属和14种．其中喜马拉雅片鳞苔福建变种（新拟名）(*Pedinolejeunea himalayensis*)（Pande & Misra）Chen et Wu var. *fukiensis* D. K. Li为新变种，台湾片鳞苔（*Pedinolejeunea formosana*（Mizt.）Chen et Wu）为福建的新记录．本文还对万木林保护区叶附生苔类植物的生态、与附主植物的关系及其地理分布关系作了探讨．
关键词：叶附生苔类；附主植物；万木林；福建

一、万木林的地理位置和生态环境

万木林自然保护区地处中国福建省的西北部，武夷山脉南侧，位于建瓯、建阳和顺昌三县交界处，即北纬27°03′，东经118°09′．其山脊最高海拔为556m，山脚海拔

*本文内的种子植物系由徐炳生教授鉴定，蕨类植物为孙吉良先生鉴定，吴鹏程教授鉴定部分苔藓标本，在此致以衷心的感谢．

300m, 林区面积2500亩．当地气候属亚热带海洋性兼内陆性气候,年平均气温18.7°C, 1月平均气温为7.5°C，7月份平均气温为28.6°C,年温差不大．年降水量为1673.3mm, 降雨多集中于春夏两季，约占全年的74%．霜期短，平均为71天．年平均相对湿度为81%．土壤以中性、微酸性红壤为主，极适宜亚热带植物的生长．万木林的林型为中亚热带常绿阔叶混交林,主要树种有壳斗科的米槠、栲树、青冈栎等, 樟科的沉水樟、闽楠、香桂, 山茶科的木荷、柃木，金缕梅科的枫树、蕈树, 杜英科的杜英、猴欢喜, 以及松科的马尾松和杉科的杉木等．下层以上述树种为主的小乔木和灌丛为叶附生苔类植物的生长提供了一个良好的生态环境．

二、万木林区叶附生苔类与附主植物的关系

表1. 万木林区叶附生苔类与附主植物的关系

Tab.1. The relationships between the epiphytes and the epiphyllous liverworts in Wanmulin Natural Reserve

Epiphyllous liverworts / Epiphytes	*Radula cavifolia*	*Radula acuminata*	*Frullania muscicola*	*Ptychanthus striatus*	*Leptolejeunea elliptica*	*Lejeunea flava*	*Lejeunea ulicina*	*Cololejeunea spinosa*	*Cololejeunea floccosa*	*Leptocolea denticulata*	*Leptocolea oblonga*	*Pedinolejeunea formosana*	*Pedinolejeunea himalayensis* var. *wuyiensis*	*Pedinolejeunea himalayensis* var. *fukiensis*
Osmunda banksiifolia		×		×	×									
Asplenium griffithianum		×												
Angiopteris fokiensis										×				
Selaginella davidii							×							
Camellia edithae	×					×						×	×	×
Maesa japonica		×						×				×		
Piper kadsura		×												
Morinda umbellata		×												
Phoebe neurantha		×			×	×						×	×	
Stauntonia chunensis											×			
Ilex ficoidea		×									×			
Prunus phaeosticta					×			×						
Chloranthus surratus		×												
Loropetalum chinense		×	×					×	×					
Setaria palmifolia		×						×	×			×		
Indocalamus tessellatas							×	×						

叶附生苔类的附主植物包括蕨类植物紫萁科的粗齿紫萁(*Omunda banksiifolia*)、铁角蕨科的厚叶铁角蕨（*Asplenium griffithianum*）、莲座蕨科的福建莲座蕨（*Angiopteris fokiensis*）和卷柏科的蔓生卷柏（*Selaginella davidii*），种子植物有山茶科的东南山茶（*Camellia edithae*）、紫金牛科的杜山（*Maesa japonica*）、胡椒科的风藤（*Piper kadsura*）、茜草科的羊角藤（*Morinda umbellata*）、樟科的白楠(*Phoebe neurantha*)，木通科的野木瓜（*Stauntonia chunensis*），冬青科的榕叶冬青（*Ilex ficoidea*），蔷薇科的腺叶野樱（*Prunus phaeosticta*），金粟兰科的及已（*Chloranthus surratus*），金缕梅科的 木（*Loropetalum chinense*）及禾本科的棕叶狗尾草（*Setaria palmifolia*）和箬竹（*Indocalamus tessellatas*）. 总计该林区附主植物有15科和16种.

上述统计表明，在万木林区着生在东南山茶及白楠植物光滑的叶面的叶附生苔类种类最多，已知有5种，其次是粗糙具毛的 木及棕叶狗尾草叶面. 它们的叶面苔类的覆盖率可达70-80%. 但不是所有植物叶面都能附生. 从大量调查表明，质地较坚硬、光滑无毛的叶面更适合多种苔类附着生长，而在较薄而柔软或粗糙有毛的叶面仅有分布较广的苔类，如尖叶薄鳞苔、尖舌扁萼苔和黄色细鳞苔等.

三、万木林叶附生苔类的地理分布关系

万木林自然保护区位于武夷山脉南侧，与江西井冈山为邻，但由于所处的地理位置、环境条件和植被构成上的差异，万木林叶附生苔类的种类组成与上述两地存在一定的差异. 在种的数目上，万木林的叶附生苔类种数少于武夷山的7科、21属及36种；而多于井冈山的2科、7属及12种. 万木林有3科、8属及14种，其中2种在武夷山尚未分布，为福建省的新记录，有7种井冈山未有分布. 显示万木林的叶附生苔类与武夷山的叶附生苔类关系较为密切，但万木林叶附生苔类垂直分布幅度较狭于武夷山和井冈山. 在武夷山，叶附生苔类见于450-1300m之间，而井冈山为650-950m，万木林区的叶附生苔类植物则见于350-450m之间. 这些特点为进一步研究万木林保护区与武夷山保护区及井冈山保护区之间的植物区系、环境条件以及叶附生苔类本身的差异提供基本的资料.

四、万木林叶附生苔类名录

Cololejeunea floccosa (Lehm. et Lindenb.) Schiffn. 18971, 18980
Cololejeunea spinosa (Horik.) Hatt. 18966
Frullania muscicola Steph. 18980
Lejeunea flava (Swartz) Nee. 18969
Lejeunea ulicina (Tayl.) Gott. 18966
Leptocolea denticulata (Horik.) Chen et Wu, 18972
Leptocolea oblonga (Herzog) Chen et Wu, 18970, 19361

Leptolejeunea elliptica (Lehm. et Lindenb.) Schiffn. 18964, 18968, 18969, 18966, 18978, 18965, 18967, 18982

Radula cavifolia Hamp. 18965

Radula acuminata Steph. 18971, 18973

Pedinolejeunea formosana (Mizutani) Chen et Wu, 18965, 18968, 18969, 18961, 18975, 18979, 18982, 19363

Pedinolejeunea himalayensis (Pande et Misra) Chen et Wu var. *wuyiensis* Chen et Wu, 18964, 18965

喜马拉雅片鳞苔福建变种 新变种 图1–7

Pedinolejeunea himalayensis (Pande et Misra) Chen et Wu var.*fukiensis* D.K. Li, var. nov. fig.1: 1–7 .

A var. *himalayensi* differt: lobulus latissimus, elongato lingulatus, 0.28 – 0.30 mm longus, 0.10 – 0.19 mm latus, complanatus, erectus, apice 1–3 cellulariter latus, basi 5–8 cellulariter latus, medio lobatus bicellularis dentatus.

Fujian: Wanmulin Natural Reserve, on the leaves of *Camellia edithae*, 26. Ⅷ, 1986, Dengke Li 18965 (Type in Herb. Mus. Nat. Hist. Shanghai)

本变种与原变种的区别在于腹瓣宽阔，尖部向内弯曲，呈长舌形，长0.28–0.36mm，宽0.10–0.19mm，扁平，尖部宽1–3细胞，基部宽3–8细胞，中部内侧具长2个细胞的齿.

福建：万木林自然保护区，东南山茶叶面，1986.8.26, 李登科18965（模式标本存上海自然博物馆）

Ptychanthus striatus (Lehm. et Lindenb.) Nee. 18978

五、讨论

1.万木林保护区内林木参天，树龄古老而郁郁葱葱，迄今已有数百年的历史. 林区外围的居民点距离较远，干扰少，自然生态系统保存良好，为叶附生苔类的生存形成有利的生境条件.

2.万木林叶附生苔类主要出现于海拔350–450m之间的沟谷林内，除安徽祁门叶附生苔类见于海拔200m外（吴鹏程、郭新弧，1980），万木林的叶附生苔类植物的垂直分布系较低海拔记录之一，其垂直幅度仅100m.

3.万木林保护区位于武夷山脉南侧，为一个低山丘陵地带. 从叶附生苔类植物的种类比较,万木林与武夷山东南坡的叶附生苔类关系较为接近. 现在万木林分布的14种叶附生苔中，与武夷山共同分布的有12种，占85.6%，但原仅分布于台湾、香港、贵州南部及西藏南部的台湾片鳞苔现见于万木林保护区. 同时，在万木林林区还发现1新变种. 1958年，陈邦杰把武夷山脉东坡以福州为界，包括广东省大部分和台湾地区，北面从五岭山脉南坡，划归为中国苔藓植被的“岭南区”，而万木林叶附生苔类种类组成与此划分相吻合.

4. 万木林保护区现知有14 种叶附生苔类生存. 该林区迄今仍然存在较好的自然植被，其西侧山脚有溪流为山谷两侧形成局部冬暖夏凉、温暖湿润的小气候，一些热带、亚热带叶附生苔类植物在这丘陵的溪边生长可能已有较长的历史.

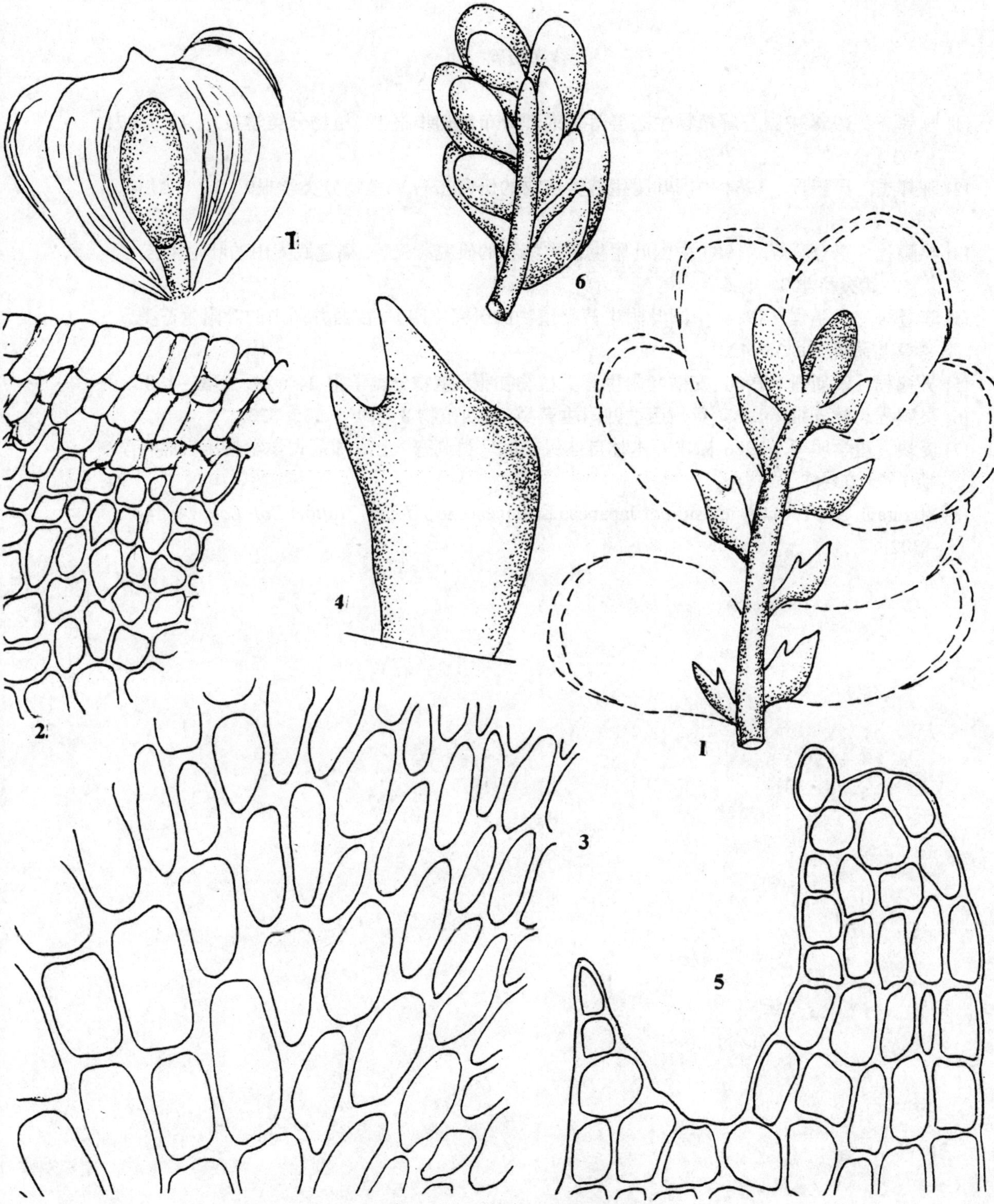

图1–7 喜马拉雅片鳞苔福建变种

Pedinolejeunea himalayensis (Pande et Misra) Chen et Wu var. *fukiensis* D.K.Li

1. 植物体一部分（腹面观）（×34）；2. 叶尖细胞（×340）；3. 叶基细胞（×340）；4. 腹瓣（×84）；5. 腹瓣（×340）；6. 成熟雄苞（×34）；7. 雌苞（腹面观）（×34）

参考资料

[1] 陈邦杰，1958 中国苔藓植物生态群落和地理分布的初步报告，植物分类学报7（4）：271 – 293.

[2] 陈邦杰、吴鹏程，1964 中国叶附生苔类植物的研究（一），植物分类学报9（3）：213–276.

[3] 吴鹏程、李登科等，1983 中国叶附生苔类植物的研究（三），福建武夷山的叶附生苔类，武夷科学 3：1–6.

[4] 李登科、吴鹏程，1988 中国叶附生苔类植物的研究（四），江西井冈山的叶附生苔类. 考察与研究 8：38–42.

[5] 吴鹏程、郭新弧，1980 安徽叶附生苔类植物初报，植物分类学报 24（2）：136–138.

[6] 吴鹏程、林齐维，1988 贵州茂兰叶附生苔类，广西植物 8（4）：335–338.

[7] 虞块、唐学明等，1986 福建万木林自然保护区鸟类调查，上海师范大学学报（自然科学版）2：53–63.

[8] Mizutani, M., 1961 A revision of Japanese Lejeuneaceae, *Journ. Hattori Bot. Lab*. 24: 146 – 302.

秦岭山脉藓类植物区系补遗

郭 伟 张满祥
(西安植物园，西安 710061)

NEW ADDITIONS TO THE BRYOFLORA OF QINLING MTS., NW CHINA

Guo Wei and Zhang Man-xiang
(Xian Botanical Garden, Xian 710061, China)

Abstract: Eight new records of mosses are reported from Qinling Mt. in southern Shaanxi Province: *Campylopus umbellatus* (Arnott.) Par., *Brachymenium ptychothecium* (Besch.) Ochi, *Pohlia gracillima* (Card.) Horik. & Ochi, *Epipterygium tozeri* (Grev.) Lindb., *Philonotis turneriana* (Schwaegr.) Mitt., *Orthotrichum macounii* Aust. ssp. *japonicum* Iwat., *Thuidium kanedae* Sak. and *Eurhynchium laxirete* Broth. ex Card. A key to the species of *Brachymenium* from Qinling Mt. is also presented.

Key Words: Mosses; *Brachymenium*; Qinling Mt.; NW China

摘要：本文报道采自陕西省南部藓类植物8个新记录种，即伞形曲柄藓*Campylopus umbellatus*、皱蒴短月藓*Brachymenium ptychothecium*、纤细丝瓜藓*Pohlia gracillima*、紫色小叶藓*Epipterygium tozeri*、东亚泽藓*Philonotis turneriana*、北美木灵藓日本亚种*Orthotrichum macounii* ssp. *japonicum*、短肋羽藓*Thuidium kanedae*和疏网美喙藓*Eurhynchium laxirete*。本文并提供秦岭山脉短月藓属*Brachymenium*的分种检索表。

关键词：新记录；藓类植物区系；秦岭山脉；中国

秦岭山脉地区藓类植物种类成分较为复杂，从北亚热带、温带到高山寒带，从湿生到旱生的区系成分在该山区均有所发现。因此，本区藓类植物曾引起不少中、外学者的浓厚兴趣。早在1896-1898年间，德国藓类植物学家C.Mueller曾先后发表了意大利人Giraldii于1893-1895年在中国陕西采集的藓类植物。1955年以后陈邦杰教授在研究中国藓类植物生态群落和地理分布时对秦岭地区藓类植物也作了重要报道。1978年张满祥发表了《秦岭植物志》第三卷第一册藓类植物。在近期调查中,笔者发现秦岭地区1个新记录属*Epipterygium* Lindb.及8个新记录种,无疑这是对《秦岭藓类志》的补充，同时，在植物地理分布上也有它特殊的意义。现作如下报道。

曲尾藓科 Dicranaceae

曲柄藓属 *Campylopus* Brid.

1.伞形曲柄藓

Campylopus umbellatus (Arn.) Par., Ind. Bryol. 264. 1894——*Thysanomitrion*

umbellatum Arn., Mem. Soc. Linn. Paris 5:263.1827——*Campylopus blumii* (Dozy & Molk.) Bosch. & Lac., Bryol. Jav. 1: 81. 1858——*Trichostomum blumii* Doz. & Molk., Ann. Sci. Nat. Bot. ser. 3,2:316. 1844——*Campylopus coreensis* Card.,Bull. Herh. Boiss. ser. 2,7:715.1907——*Campylopus coreensis* var. *amoyensis* Dix. & Ther., Hong Kong Naturalist Suppl. 2:5, 1933——*Campylopus dozyanus* (C.Muell.) Jaeg., Ber. Thatigk. St. Gall.Naturw.Ges.1870 -71:418,1872——*Campylopus dozyi* C. Muell. ex Kindb., Enum. Bryin.Exot.50, 1888——*Campylopus ferroei* Broth. in Par., Index Bryol. Suppl. 92.1900——*Campylopus leptoneuron* Broth. ex Ihs., Cat. Moss Japan 32,1929——*Campylopus nagasakinus* Broth. in Par., Index Bryol. Suppl. 94. 1900——*Campylopus viridulus* Card.,Bull. Herb. Boiss. ser.2,7:715,1907——*Campylopus scabripilus* Warnst., Hedwigia 57:80,1915——*Campylopus richardii* auct. non Brid.

陕西：城固县，天明区，九坝，张满祥等2666、2644、2673、2649；同地，五堵门区，元山，张满祥、张继祖2388、2392、2426.

生于海拔560－670米处向阳干燥岩面薄土或林下砂土上．植株大形，密集丛生呈垫状．在该区5月底到6月初常可见有多数孢子体丛生于枝顶或新枝腋部，但蒴柄尚未伸出，孢蒴幼嫩尚未发育成熟．

本种过去仅知产于福建、广东、海南、台湾、安徽、江西、江苏、浙江、湖南、湖北、云南、贵州、四川和西藏；以及日本、朝鲜、印度尼西亚和大洋洲的热带和亚热带广阔地区．在中国秦岭山脉为新记录．

真藓科 Bryaceae

短月藓属 *Brachymenium* Schwaegr.

2.皱蒴短月藓

Brachymenium ptychothecium (Besch.) Ochi,in Ady. Front.Pl.Sci.4:108.1963——*Bryum ptychothecium* Besch., Ann. Sci. Nat. Bot. ser. 7.15:66.1892；西藏苔藓植物志，169页，1985.

陕西：城固县，五堵门区，元山，张满祥、张继祖2407e.

生于海拔600m处青麸杨树干上．本种原仅知产于四川、云南和西藏；以及印度、锡金及尼泊尔．在秦岭山脉为新记录．

短月藓属植物多分布暖热地区．秦岭地区现知有该属的3种,现将它们区分如下:

秦岭地区短月藓属种的检索表

1.植物体形小，土生；叶无分化边缘．仅见于陕西省城固县天明区……………… 1.纤枝短月藓*Brachymenium exile*

1.植物体形大，树生；叶具分化边缘，………………………………………… 2.

2.植株深绿色，不具光泽．叶片长卵形，分化边较薄；中肋粗壮、突出叶尖而呈长芒状．分布秦岭各地…………………………………………………………

2. 尼泊尔短月藓*Brachymenium nepalense*

2.植株淡绿色，具光泽．叶卵状披针形，分化边较厚；中肋略突出于叶尖．仅见于陕西城固县五堵门区元山………………………………………………………3.

皱蒴短月藓*Brachymenium ptychothecium*

丝瓜藓属 *Pohlia* Hedw.

3.纤细丝瓜藓

Pohlia gracillima (Card.) Horik. et Ochi, Lib. Arts Journ.Tottori Univ. Nat. Sci. 4:13,1954——*Webera gracillima* Card.,Bull.Soc.Bot.Geneveser.2,1:125.1909; 西藏苔藓植物志，173页，1985.

陕西：城固县，天明区，九坝，张满祥等2555、2572.

生于海拔550m处杂木林下路旁土壁或砂土上．据文献记载本种分布于云南和西藏；以及南亚、亚洲东部．现在秦岭山脉为新记录．

多产于亚热带和热带地区．茎纤细．叶片排列稀疏，狭长披针形，干时强烈卷扭；细胞壁较薄．无性芽长棒形，多生于叶尖或叶背面．

小叶藓属 *Epipterygium* Lindb.

4.紫色小叶藓

Epipterygium tozeri (Grev.) Lindb., Oefv. Svensk. Vet. Akad. Foerh. 21: 576,1865; J.Shaw in Bryologist 87:136,1984; Iwatsuki & M. Mizutani in coloured Illustrations of Bryophytes of Japan 111.1981——*Webera tozeri* (Grev.) Schimp. Coroll. Bryol. Eur. 67.1856——*Bryum tozeri* Grev. Scott. Crypt.F1. 5:285.1827——*Anisostichium tozeri* (Grev.) Mitt., J. Linn. Soc. Bot.7:119. 1864——*Pohlia tozeri* (Grev.) Delogn., Fl. Crypt. Belg. 159,1884.

陕西：城固县，天明区，九坝，张满祥等2548、2569和2618.

生于海拔550－570米处松、栎林下土上或土坡．在该区常与睫毛苔*Blepharostoma trichophyllum*组成群落．本种以往记录于河北、四川、云南、浙江、西藏和台湾；以及日本、印度尼西亚、高加索地区和北美等地．属和种均为首次记录于秦岭山脉．

为小形土生藓类．在不孕枝上叶排列呈3－4行(侧叶2列、背叶1－2列),背、侧叶异形，而与真藓科其它属植物相区别．

珠藓科 Bartramiaceae

泽藓属 *Philonotis* Brid.

5.东亚泽藓

Philonotis turneriana (Schwaegr.) Mitt., Journ. Linn. Soc. Bot. Suppl.1:62, 1859——*Bartramia turneriana* Schwaegr., Spec. Musc. Suppl.3(1):238,1828;西藏苔藓植物志，224页，1985.

陕西：城固县，天明区，九坝，张满祥等2697、2566、2538、2736、2637、2665a、2574、2540、2728和2767；同县，二里区，大同村，张满祥等2837、2721、

2821; 同县, 五堵门区, 元山, 张满祥和张继祖2393、2395、2404、2364a、2484c、2453、2443、2367、2446a、2380a、2365、2442、2369a、2486、2499.

生于海拔570–680m处林下向阳潮湿土坡、滴水岩缝及小溪边砂土上. 中国台湾、福建、广东、广西、江苏、江西、安徽、四川、云南和西藏; 以及日本、印度尼西亚和尼泊尔有记录. 在秦岭山脉为新记录.

该种形态变异较大, 在秦岭山区标本叶中肋仅略突出叶尖, 与原种描述中肋突出叶尖较长相不一致, 而其它特征与原种描述相同.

木灵藓科 Orthotrichaceae

木灵藓属 *Orthotrichum* Hedw.

6.北美木灵藓日本亚种

Orthotrichum macounii Aust. ssp. *japonicum* Iwatsuki, Journ. Hattori Bot. Lab. 21:240, 1959;西藏苔藓植物志, 235页, 1985.

陕西: 城固县,五堵门区,元山,张满祥和张继祖2431b、2355、2687、2476b、2430d和2523d.

生于海拔600–680m处青麸杨树干上. 该种原记录于西藏以及日本. 在秦岭山脉为新记录种.

羽藓科 Thuidiaceae

羽藓属 *Thuidium* B.S..G.

7.短肋羽藓

Thuidium kanedae Sak., Bot. Mag. Tokyo 57:345.3,1943; Iwatsuki & M.Mizutani in Coloured Illustrations of Bryophytes of Japan 200. 1981.

陕西: 城固县, 五堵门区, 元山, 张满祥和张继祖2514a、2506c、2350、2522b、2403b、2530d; 同县, 小河区, 砖溪乡, 张满祥、张继祖1349; 同县, 天明区, 九坝, 张满祥等2611、2552、2785; 同县, 高坝,张满祥等2573a、2568、2780、2709; 同县, 二里区, 毛垭村, 张满祥等2664、2674、2565、2541、2658; 同县, 大同村, 张满祥等2819.

生于海拔600–880m处松栎林下岩面薄土、砂土和青麸杨树干. 分布河北、江苏、浙江、安徽、湖南、四川和台湾; 以及日本和朝鲜. 首次记录于秦岭地区, 为东亚特有种.

本种外形和叶形与大羽藓*Thuidium cymbifolium*非常相似, 不同点是茎叶中肋于叶中上部即消失、茎叶细胞常具2一4个粗疣.

青藓科 Brachytheciaceae

美喙藓属 *Eurhynchium* B.S.G.

8.疏网美喙藓

Eurhynchium laxirete Broth. ex Card., Bull. Soc. Bot. Geneve ser.2,4:380, 1912.; 西藏苔藓植物志, 375页, 1985.

陕西：城固县，天明区，高坝，张满祥等2717.

生于海拔650m处栎林下土上. 曾记录于福建、台湾、安徽、江苏、江西和西藏；以及日本和朝鲜. 在秦岭地区为新记录.

致谢：吴鹏程教授赐寄宝贵对照标本，黎兴江教授帮助查阅有关资料和查阅修改全文，张继祖教授参加全部野外工作，房宁君和王明先生参加过部份野外工作，仅此深表感谢.

参考文献

[1] 陈邦杰，1958. 中国苔藓植物生态群落和地理分布的初步报告，植物分类学报7(4): 271–293.

[2] 黎兴江等，1985:西藏苔藓植物志1－581页，科学出版社.

[3] 张满祥，1978，秦岭植物志三卷一册:苔藓植物门，1－329页. 科学出版社.

[4] 吴鹏程、罗健馨、汪楣芝，1984: 苔藓名词及名称，1－124页，科学出版社.

[5] 胡晓云、吴鹏程，1991: 四川金佛山藓类植物区系的研究，植物分类学报29(4): 315–334.

[6] Brotherus, V. F., 1909: In A, Engler and K. Prantl, Die Naturlichen Pflanzenfamilien l. 3.554–555. Engelmann. Leipzig.

[7] Chuang, C. -C., 1973. A moss flora of Taiwan excilusive of essentially pleurocarous families. *J. Hattori Bot. Lab.* 37:419–509.

[8] Frahm, J-P. 1992. A Revision of the East-Asian species of *Campylopus. J.Hattori Bot.* 71:133–164.

[9] Hu Renling et Wang Youfang, 1987. A Review of the Moss Flora of East China. Memoirs,New York Bot. Gard. 45:455–465.

[10] Iwatsuki, Z.,and M.Mizutani, 1981: Coloured Illustrations of Bryophytes of Japan (in Japaneses).

[11]Lin Shanhsiung, 1988. List of Mosses of Taiwan. *Yushania* 5(4):1–40.

[12]Mueller, C., 1896-98. Bryologia Provinciae Schen-si Sinensis.Nuov.Giornal Botanica Italiano n.111–V.

[13]Redfearn, P.L. et P. C. Wu, 1986. Cataleg of Mosses of China. Ann. Missouri Bot. Card. 73(1):177–208.

[14]Schaw, Jonathan, 1984. Quantitative Taxonomic Study of Morphology in *Epipterygium. The Bryologist* 87(2): 132–142.

[15]Zhang Manxiang, 1987. The moss plant communities in the Qingling Range of China. Symp. Biol. Hungar 35:497–513.

[16] ————————, 1984. Preliminary studies in the floristic characteristics and geographic distribution of mosses in the Qinling Range. *J.Hattori Bot. Lab.* 56: 39–43.

[illegible]

[illegible]

参考文献

[1] [illegible]

[2] [illegible]

[3] [illegible]

[4] [illegible]

[5] [illegible]

[6] Brotherus V.F. [illegible] Natürlichen Pflanzenfamilien [illegible]

[7] [illegible] 1957. A [illegible]

[8] [illegible]

[9] [illegible] Moss Flora of [illegible]

[10] [illegible] Coloured illustrations of Bryophytes of Japan [illegible]

[11] [illegible]

[12] [illegible]

[13] [illegible]

[14] [illegible]

[15] [illegible] The moss plant communities of the [illegible]

[16] [illegible]

江苏省苔藓植物研究 I *

凌萍萍　　　汪　庆
(中国科学院江苏植物研究所，南京210014)

THE STUDY OF BRYOPHYTE FLORA OF JIANGSU PROVINCE I

Lin Ping-ping and Wang Qing
(*Jiangsu Institute of Botany,Jiangsu Province,
Academia Sinica, Nanjing 210014, China*)

Abstract: This is the first of a series of papers designed to study the bryophytes of Jiangsu Province. A review of past bryological research activities conducted in Jiangsu Province and the geology and natural history of the province were presented. At present, the checklist of Jiangsu bryophytes consists 283 species in 61 families, of which 235 are moss species belonging to 38 families.
Key words: Jiangsu Province; Bryophytes; Mosses

摘要：本文是江苏省苔藓植物研究报告之一，简要介绍了江苏省自然地理状况，苔藓植物调查与研究历史．江苏藓类植物名录报导了藓类植物38科和235种．
关键词：江苏省；苔藓植物；名录

一、江苏省自然地理概况

江苏省位于中国东部，北接山东省，南连上海市和浙江省，西邻安徽省，东滨黄海，地居淮河、长江下游，界于北纬35°07'-30°46'，东经116°22'-121°55'，南北跨度4°11'，最人直线距离为460多公里，东西间经度差为5°33'．总面积为10万2千100平方公里．

江苏省的地层和地质构造，可以响水、清江、盱眙一线分为南北两区．北区同我国华北地区的地层和地质构造基本一致，形成于太古代，构造比较稳定，是华北古陆（或准地台）的组成部分；南区同云南省东北部以东的长江流域的地质和地层构造基本一致，形成于元古代，构造不如北区稳定，是扬子古陆（或准地台）的组成部分．在地层层南区，自震旦纪以来，各地质时代的地层发育完整；北区古生代和中生代的地层则有部分缺失．

江苏平原以面积辽阔、地势低平、河网稠密、湖荡众多为特征，也有少数低山丘陵和岗地分布在它的东北部和西南部．在东北部有连云港的云台山、赣榆、东海与山东省交界处的一些低山丘陵．在西南部有盱眙、六合、仪征间的许多方山丘陵、江浦

* 江苏省自然科学基金项目
This project is supported by the Jiangsu Natural Science Foundation.

的老山山脉、南京镇江间的宁镇山脉、句容溧水、金坛间的茅山山脉、宜兴溧阳南部的宜溧山脉等．所有这些低山丘陵都以山体破碎、山势和缓低矮为特征，海拔高度多在200-300m之间，只有少数山峰突起在400-500m，甚至600m以上，其中以云台山望海楼最高，海拔625m．此外在徐州市郊尚有少数石灰岩残丘，在长江三角洲上也散布着一系列孤山残丘．

江苏省跨暖温带、北亚热带和中亚热带三个自然带，年平均温13.5-15.5°C，最冷月均温-0.3-3°C，最热月均温26-28°C，年平均降水量700-1200毫米以上．一般说来，苏北灌溉总渠以北的徐淮地区属于与暖温季风气候相适应的落叶阔叶林，由此向南逐渐递变为与北亚热带季风气候相适应的落叶阔叶与常绿阔叶混交林类型，到最南部的宜溧山地再递变为与中亚热带季风气候相适应的常绿阔叶林类型．地带性土壤依次为棕壤、淋溶褐土、黄棕壤和黄壤；非地带性土壤包括滨海平原盐土、沿湖和沿江低地沼泽土、冲积平原草甸土和花碱土．

二、江苏省苔藓植物研究简史

江苏省苔藓植物区系资料的积累始于1891年，E.Faber在南京采集并由E.S.Salmon报道的*Pseudoleskea capillata* Besch.．在1906年至1922年期间，法国传教士Rev. P. Courtois和A.Henry分别在江苏、安徽、江西和浙江等地进行采集．Paris在研究整理上述苔藓植物后，报导了江苏苔藓17种，其中藓类16种和苔类1种，包括模式产地在南京、无锡、苏州的新种六个：*Entodon henryi* Par. et Broth.、*Schwetschkea courtoisii* Par. et Broth.、*Ctenidium robusticaule* Broth. et Par.、*Fissidens nankingensis* Broth. et Par.、*Macromitrium ousiense* Broth. et Par. 和 *Leskea subfiliramea* Broth. et Par. 在Paris故后所余标本由Potier de la Varde继续研究．1929年，H.N.Dixon发表了Rev. Pere E. Licent在南京采集的一种*Hypnum alare*（C.M.）Par.．其后陈邦杰(1933－1934)在宜兴等地进行采集．1958年，胡人亮对吴县洞庭西山苔藓植物进行了研究，记录了苔藓植物70余种．1978年吴鹏程、金岳杏在宜兴、无锡、吴县进行采集．1982至1984年李登科、高彩华在宜兴、吴县常热、昆山进行采集，据此在《长江三角洲及邻近地区孢子植物志》中报道江苏苔藓．本文作者们自1991年起在全省范围内进行广泛采集，共获得标本3440份，并同馆藏的吴鹏程、金岳杏在宜兴、无锡、吴县采集标本871份进行整理鉴定，以及北京中国科学院植物所、上海自然博物馆所藏江苏苔藓标本进行整理．现知，江苏省共有苔藓植物283种，隶属于61科，其中藓类植物为38科及235种．

三、江苏省苔藓植物名录

I .藓纲 Musci

Sphagaceae 泥炭藓科

1. *Sphagum palustre* L.世界广布成分

2. *Sphagum microporum* Warnst.et Card.中国一日本成分
吴县西山、溧水迴峰山、费达无号；宜兴芙蓉山、单人骅等无号；宜兴林场、吴鹏程 、金岳杏2410.

Archidiaceae 无轴藓科

3. *Archidium* ohioense Schimp.ex C.Muell.东亚一北美 间断成分
吴县东山、刘仲苓32480；宜兴龙池、秦怀兰183.
4. *Archidium alternifolium* (Dicks.) Schimp.
南京方山、臧穆、方2.

Ditrichaceae 牛毛藓科

5. *Ceratodon purpurense* (Hedw.) Brid.世界广布成分
宜兴老鹰界，凌萍萍，1647.
6. *Ditrichum divaricatum* Mitt. 中国一日本成分
昆山亭林，凌萍萍、汪庆，3315.
7. *Ditrichum heteromallum* (Hedw.) Brit.世界广布成分
南京老山，凌萍萍1630；句容茅山，凌萍萍041，074；宜兴老鹰界，凌萍萍，汪庆2071；连云港宿城，凌萍萍156，177，394，397，405，441；花果山，凌萍萍299，472；柳河，凌萍萍332；鹰登山，凌萍萍463.
8. *Ditrichum pallidum* (Hedw.) Hamp.北温带成分
句容茅山，凌萍萍056，119，117，090，079，073；南京紫金山，凌萍萍011；无锡惠山，凌萍萍、汪庆2267，2269；马山，凌萍萍951；宜兴馨山，吴鹏程；金岳杏 2478；南京宝华山，凌萍萍1530，1520；吴县东山，吴鹏程、金岳杏1669；吴县灵岩山，吴鹏程、金岳杏1881；吴县西山，吴鹏程、金岳杏1790；吴县光福，凌萍萍、汪庆2377；凌萍萍719, 869；灵岩山, 凌萍萍559；东山，凌萍萍756，582；连云港花果山，凌萍萍、汪庆32346；连岛，凌萍萍、汪庆3072；大桅尖，凌萍萍131，185，377，399.

Dicranaceae曲尾藓科

9. *Trematodon longicollis* Michx.世界广布成分
吴县灵岩山，刘仲苓32562；溧阳平桥，李登科18912；南京栖霞山，秦怀兰385；中山植物园，凌萍萍931，935.
10. *Dicranella heteromalla* (Hedw.) Schimp.世界广布成分
宜兴善卷洞李登科00256；常熟虞山，凌萍萍、汪庆2222；宜兴小黑沟，凌萍萍、汪庆2135；邓懋彬7102a,7095；宜兴龙池，吴鹏程、金岳杏2287；宜兴馨山，吴鹏程、金岳杏2448；吴县西山，吴鹏程、金岳杏1829；吴县东山，凌萍萍、汪庆2511；吴县光福，凌萍萍826，841，869；苏州穹窿山，凌萍萍678；无锡马山，凌萍萍、汪庆3099；柳河凌萍萍、汪庆3240.
11. *Dicranella coarctata* (C.Muell) Bosch. et Lac. 北温带成分

常熟虞山，李登科17792；吴县东山，凌萍萍、汪庆2514；无锡马山，凌萍萍991；连云港、连岛，凌萍萍、汪庆3099，3072；南京宝华山，凌萍萍1550；连云港花果山，凌萍萍、汪庆3215.

12. *Dicranodontium asperulum* (Mitt.) Broth. 北温带成分
常熟虞山，李登科17595；镇江南山，凌萍萍、汪庆2742；常熟虞山，凌萍萍、汪庆2215，2218，2256；无锡惠山、凌萍萍、汪庆2270；扬州何园2855.

13. *Dicranodontium denudatum* (Bird.) Broth. 北温带成分
吴县灵岩山，刘仲苓032559；吴县光福，凌萍萍、汪庆2433；句容茅山，凌萍萍066.

14. *Oncophorus wahlenbergii* Bird. 北温带成分
常熟虞山李登科017839.

15. *Dicranum nipponense* Besch.北温带成分
常熟虞山，李登科017553；吴县西洞庭山，吴鹏程、金岳杏1751，1760，1803，1853，1761.

16. *Dicranum scopellifolium* C.Muell.
《华东五省一市植物名录》

17. *Dicranum scoparium* Hedw.北温带成分
常熟虞山，李登科017656；宜兴龙池，吴鹏程、金岳杏2259；吴县西山，吴鹏程金岳杏1802，1708；无锡马山，凌萍萍986，1020；吴县穹窿山，凌萍萍665.

18. *Campylopus umbellatus* (Arnuth.) Par.---C. richardii Brid.
常熟虞山，李登科017849；凌萍萍、汪庆2238；宜兴龙池，吴鹏程、金岳杏2278；吴县穹窿山，凌萍萍840.

19. *Campylopus atro-virens* De Not.北温带成分
常熟虞山，李登科17540；南通军山，凌萍萍、汪庆2911，连云港柳河，凌萍萍350；连云港大桅尖，凌萍萍、汪庆3113；吴县穹窿山，凌萍萍1058，645；白沙山，凌萍萍835；灵岩山，凌萍萍832.

Leucobryaceae白发藓科

20. *Leucobryum neilgherrense* C.Muell. 热带亚洲及太平洋成分
吴县东山，凌萍萍573，745；穹窿山，凌萍萍658；句容茅山，凌萍萍112；连云港连云中学附近，凌萍萍496，497；大桅尖至宿城途中，凌萍萍、汪庆3127；无锡鼋头渚，凌萍萍、汪庆2297；南通军山，凌萍萍、汪庆2880，2890；宜兴小黑沟，凌萍萍、汪庆2127；龙池，凌萍萍、汪庆2187；常熟虞山，凌萍萍、汪庆2221；吴县西山，吴鹏程、金岳杏1770，1809.

Fissidentaceae 凤尾藓科

21. *Fissidens bryoides* Hedw.北温带成分

宜兴林场，凌萍萍、汪庆2082；苏州虎丘，林邦娟2；南京中山植物园，凌萍萍1680，1666；扬州大明寺，凌萍萍、汪庆2812；句容茅山，凌萍萍、汪庆2305，2309；镇江南山凌萍萍、汪庆2747，2749；镇江焦山，凌萍萍、汪庆2656，2660；南通狼山，凌萍萍、汪庆2922，2926；宜兴善卷洞，凌萍萍、汪庆2170.

22. *Fissidens adelphinus* Besch.中国-日本成分

吴县西山，吴鹏程、金岳杏1856；凌萍萍847；白沙山，凌萍萍583；天平山，凌萍萍、汪庆2602；灵岩山，李登科18832；南京中山植物园，凌萍萍1657；宜兴小黑沟，凌萍萍、汪庆2193；句容茅山，凌萍萍、汪庆2319；连云港柳河，凌萍萍324;花果山，凌萍萍、汪庆3234，宿城，凌萍萍、汪庆3181a；后云台山，凌萍萍、汪庆2973；黄窝，凌萍萍、汪庆3041；无锡鼋头渚，凌萍萍、汪庆2298；常熟虞山，凌萍萍、汪庆2245，2247.

23. *Fissident nobilis* Griff.热带亚洲成分

宜兴茗岭，吴鹏程、金岳杏2298；老鹰界，吴鹏程、金岳杏2189，2161，2199，2229；馨山，吴鹏程、金岳杏2482，2473；常熟虞山，李登科017793；吴县东山，凌萍萍、汪庆2503；南京中山陵，凌萍萍1589.

24. *Fissidens cristatus* Wils. ex Mitt.世界广布成分

宜兴太华，吴鹏程、金岳杏2057，张公洞、吴鹏程、金岳杏2431，老鹰界，凌萍萍、汪庆2052-b，宜兴林场，凌汪2072，小黑沟，凌萍萍、汪庆2123；吴县西山，吴鹏程、金岳杏1861,1864；灵岩山,吴鹏程、金岳杏1889,1906；无锡市植物园，凌萍萍965；常熟虞山，李登科017556；南京中山植物园，凌萍萍1607;老山滴水珠，凌萍萍1629；连云港连岛，凌萍萍、汪庆3095；黄窝，凌萍萍、汪庆3034；柳河，凌萍萍223.

25. *Fissidens geminiflorus* Doz. et Molk. var. *nagasakinus* (Besch.) Wats.
中国一日本成分

连云港水帘洞，凌萍萍216；黄窝，凌萍萍、汪庆3023；宿城,凌萍萍、汪庆3171;柳河，凌萍萍、汪庆3146；万寿涯，凌萍萍203；宜兴善卷洞，李登科0233.

26. *Fissidens perdecurrens* Besch.中国-日本成分

连云港黄窝，凌萍萍、汪庆3044；大桅尖至宿城，凌萍萍、汪庆3110；宿城，凌萍萍、汪庆3202，3203.

27. *Fissidens gymnogynus* Besch.

连云港后云台山，凌萍萍、汪庆2973；望海楼，凌萍萍132；柳河，凌萍萍291.

28. *Fissidens plagiochiloides* Besch.

宜兴善卷洞，吴鹏程、金岳杏2001；老鹰界，吴鹏程、金岳杏2244.

29. *Fissidens taxifolius* Hedw.世界广布成分

南京紫金山，凌萍萍938，老山滴水珠，凌萍萍1656；宜兴林场，凌萍萍、汪庆2068，2081；连云港，凌萍萍、汪庆3239，黄窝，凌萍萍、汪庆3027；柳河，凌萍萍335.

30. *Fissidens zippelianus* Doz.et Molk.旧世界热带成分
苏州香雪海，凌萍萍847；吴县东洞庭山，吴鹏程、金岳杏1711，1719.

31. *Fissidens zollingeri* Mont.
吴县灵宫殿，胡人亮18.

Pottiaceae丛藓科

32. *Gymnostomum calcareum* Nee.et Hornsch.
扬州瘦西湖，凌萍萍、汪庆2799；个园，凌萍萍、汪庆2849；南通狼山，凌萍萍汪庆2920;徐州铜山,凌萍萍、汪庆295；句容宝华山，凌萍萍1515；苏州虎丘，万宗玲41；南京栖霞山，凌萍萍048.

33. *Tortella tortuosa* (L.) Limpr. 世界广布成分
南京明孝陵，臧穆5601；吴县，刘仲苓32550.

34. *Pseudosymblepharis angustata* (Mitt.) Hilp.
宜兴太华，吴鹏程、金岳杏2062.

35. *Weissia crispa* (Hedw.) Mitt.
镇江焦山，凌萍萍、汪庆2654；吴江黎里，凌萍萍、汪庆2945；句容宝华山，凌萍萍1522，1516；吴县西山，李登科8820.

36. *Weissia adentula* Mitt.
徐州泉山，凌萍萍、汪庆2961；铜山，凌萍萍、汪庆2954@；常熟虞山，李登科，17535；吴县西山，李登科18760；江宁汤山，万宗玲63-13.

37. *Weissia controversa* Hedw.世界广布成分
镇江焦山，凌萍萍、汪庆2681，2655；南山，凌萍萍、汪庆2720，2733；扬州瘦西湖,凌萍萍、汪庆2741；吴江黎里2349；海门永隆沙2943;南通黄泥山，凌萍萍、汪庆2860；南京中山植物园，凌萍萍；连云港连岛，凌萍萍、汪庆3053；宿城，凌萍萍、汪庆3165；徐州汉王乡，凌萍萍、汪庆2955；宜兴湖汶东岭，邓懋彬无号；吴县东山，李登科019467.

38. *Weissia semipallida* C.Muell.
扬州瘦西湖，凌萍萍、汪庆2795，2781；吴县东山，李登科6726.

39. *Weissia exserta* (Broth.) Chen
镇江南山，凌萍萍、汪庆2739.

40. *Oxystegus cylindricus* (Brid.) Hilp.
苏州虎丘，李兆兰17；南京明孝陵，叶德娴4.

41. *Trichostomum crispulum* Bruch北温带成分
句容宝华山，凌萍萍1540；宜兴善卷洞，李登科240，秦怀兰252；吴县西山，李登科6767.

42. *Hyophila involuta* (Hook.) Jaeg.
镇江南山，凌萍萍、汪庆2727，2762；扬州瘦西湖2797;吴县灵岩山,李登科6867.

43. *Hyophila propagulifera* Broth.
镇江焦山，凌萍萍、汪庆2394；金山，凌萍萍、汪庆2707；南山，凌萍萍、汪庆2737；南通狼山，凌萍萍、汪庆2917；南京灵谷寺，李登科16562.

44. *Barbula indica* (Hook.) Spreng.
宜兴善卷洞，李登科252.

45. *Barbula unguiculata* Hedw.
扬州太明寺，凌萍萍、汪庆2327；句容茅山，凌萍萍、汪庆2304；海门永隆沙，凌萍萍、汪庆2938；南京植物园，凌汪335；宜兴善卷洞，李登科234；吴县东山，凌萍萍545，茅蓬保护区，李登科19490.

46. *Didymodon rigidicola* (Broth.) Zander
宜兴龙池，吴鹏程、金岳杏2317.

47. *Didymodon vinealis* (Bird.) Zander
吴县灵岩寺，凌萍萍、汪庆2641；宜兴林场，凌萍萍、汪庆2109；句容宝华山，凌萍萍1514.

48. *Scopelophila cataractae* (Mitt.) Broth.
吴江黎里，凌萍萍、汪庆2947，2948；常熟，李登科17460.

49. *Pottia lanceolata* (Hedw.) C.Muell.
常熟虞山，李登科17548，17590.

50. *Pottia intermedia* (Turn.) Fuernr.
溧阳松岭，李登科18953.

51. *Tortula muralis* Hedw.
苏州虎丘，凌萍萍、汪庆2585；拙政园，凌萍萍、汪庆2548；吴江黎里，凌萍萍汪庆2946；南京紫金山，凌萍萍07；吴县西山，李登科6779.

Ptychomitriaceae缩叶藓科

52. *Ptychomitrium dentatum* (Mitt.) Jaeg.
南京中山植物园，凌泮萍、汪庆3431.

53. *Ptychomitrium linearifolium* Reim.et Sak.中国-日本成分
宜兴丁山，李登科217；南京，顾其敏149；宜兴老鹰界，凌萍萍、汪庆2134；连云港柳河，凌萍萍、汪庆3152；大桅尖，凌萍萍、汪庆3138；花果山，凌萍萍259,298.

54. *Ptychomitrium sinense* (Mitt.) Jaeg.中国-日本成分
常熟虞山，李登科017611；吴县东山，李登科6741；句容茅山，李登科020295,凌萍萍57,866，103a;宜兴老鹰界，吴鹏程、金岳杏2231；凌萍萍、汪庆2149；昆山亭林，凌萍萍、汪庆3284；吴县灵岩山，吴鹏程、金岳杏1898，1884；凌萍萍829；东山，凌萍萍764，拙政园738；虎丘，737，狮子林，凌萍萍、汪庆2540，2544；南通狼山，凌萍萍、汪庆2930；扬州个园，凌萍萍、汪庆2846；连云港花果山，凌萍萍、汪庆3235；大桅尖，凌萍萍、汪庆3140，3117；南京中山植物园，凌萍萍、汪庆3425.

55. *Ptychomitrium wilsonii* Sull. et Lesq.中国–日本成分
溧阳松岭，李登科17611；宜兴小黑沟，李登科20398，龙池，秦怀兰93，吴鹏程，金岳杏2280，小黑沟，吴鹏程、金岳杏2488，馨山，吴鹏程、金岳杏2459；常熟虞山，李登科7803.

Grimmiaceae紫萼藓科

56. *Grimmia laevigata* (Brid.) Brid. 世界广布成分
吴县灵岩山，李登科18810；常熟虞山，李登科17843；宜兴龙池，吴鹏程、金岳杏2279，2327，2330；老鹰界，吴鹏程、金岳杏2196.

57. *Grimmia affinis* Hornsch.
宜兴乾元界，吴鹏程、金岳杏2126；句容茅山，凌萍萍、汪庆2361，2347；小黑沟，凌萍萍、汪庆2141；吴县灵岩山，凌萍萍、汪庆2542，吴鹏程、金岳杏1903；南通狼山，凌萍萍、汪庆2929；连云港大桅尖，凌萍萍、汪庆2998；花果山，凌萍萍、汪庆3227；宿城，凌萍萍、汪庆3114；柳河，凌萍萍、汪庆3149.

58. *Grimmia pilifera* P.Beauv. 北温带成分
南京灵谷寺植三第三组96959；溧阳横涧，李登科18891；句容茅山，凌萍萍072，089，凌萍萍、汪庆2360，2355；宜兴老鹰界，吴鹏程、金岳杏2178；吴县灵岩山，凌萍萍565，吴鹏程、金岳杏1887；吴县西山，吴鹏程、金岳杏1783，1821;连云港大桅尖，凌萍萍、汪庆2977；黄窝，凌萍萍、汪庆3038；南京中山植物园，凌萍萍1653；吴县天平山，陈邦杰120.

59. *Grimmia decipiens* (Schultz.) Lindb.
南京中山植物园，凌萍萍、汪庆3427；常熟虞山，凌萍萍、汪庆2239；宜兴龙池，凌萍萍、汪庆2195，2016.

60. *Grimmia tergestina* Tomm. ex B.S.G.
昆山亭林，凌萍萍、汪庆3339.

61. *Rhacomitrium canescens* (Hedw.) Brid.世界广布成分
句容茅山，凌萍萍110；连云港宿城，凌萍萍、汪庆3163；望海楼，凌萍萍158.

62. *Rhacomitrium heterostichum* (Hedw.) Brid.世界广布成分
吴县西山，凌萍萍、汪庆2473，吴鹏程、金岳杏1786；穹窿山，凌萍萍663；灵岩山，凌萍萍856，吴鹏程、金岳杏1894；东山，凌萍萍303；无锡鼋头渚，吴鹏程、金岳杏1937；马山，凌萍萍976，1027.

63. *Rhacomitrium anomodontoides* Card.
句容茅山，凌萍萍083，凌萍萍、汪庆2350；吴县光福，凌萍萍、汪庆2440.

Funariaceae葫芦藓科

64. *Physcomitrium sphaericum* (Hedw.) Fuernr.
连云港连岛，凌萍萍、汪庆3065；大桅尖，凌萍萍518；南京中山植物园,邓懋彬无号.

65. *Physcomitrium eurystomum* Sendtn.

南京师大校园，秦怀兰12；吴县东山，凌萍萍、汪庆2519；光福，凌萍萍569；常熟，凌萍萍、汪庆2214.

66. *Physcomitrium japonicum* (Hedw.) Mitt.

67. *Physcomitrium courtoisi* Par. et Broth.

南京紫金山，凌萍萍06，017；句容茅山，凌萍萍、汪庆2324.

68. *Funaria hygrometrica* Hedw.世界广布成分

射阳，生于近人处有机质丰富的土地上，刘方熏7154.

Bryaceae真藓科

69. *Brachymenium nepalense* Hook.中国-日本成分

宜兴林场，凌萍萍、汪庆2086；苏州留园，凌萍萍、汪庆2563；扬州何园，凌萍萍、汪庆2850.

70. *Bryum argenteum* Hedw. 世界广布成分

南京植物园，凌萍萍1617；老山，凌萍萍1682；句容茅山，凌萍萍、汪庆2332；无锡惠山，凌萍萍、汪庆2271；无锡蠡园，吴鹏程、金岳杏1981；镇江南山，凌萍萍、汪庆2740；南通狼山，凌萍萍、汪庆2931；黄泥山，凌萍萍、汪庆2859；昆山亭林，凌萍萍、汪庆3299，3274；徐州铜山2956；连云港玉女峰，凌萍萍863；吴县东山，吴鹏程、金岳杏1629，凌萍萍655；光福，凌萍萍560，56,862；穹窿山，凌萍萍724；苏州虎丘，凌萍萍758.

71. *Bryum atrovirens* Will. ex Brid.

宜兴龙池，凌萍萍、汪庆2131，2132；老鹰界2974；

72. *Bryum caespiticium* L.ex Hedw. 世界广布成分

南京植物园，凌萍萍1663，907；汤山、黎兴江12；宜兴老鹰界，凌萍萍、汪庆2085；苏州虎丘，凌萍萍757，773；徐州铜山，凌萍萍、汪庆2952；连云港花果山，凌萍萍、汪庆3221；大桅尖，凌萍萍455，凌萍萍、汪庆2975；黄窝3099.

73. *Bryum capillare* L. ex Hedw. 世界广布成分

南京植物园，凌萍萍1641；宝华山，凌萍萍1528；昆山亭林，凌萍萍、汪庆3278，3266；常熟虞山，凌萍萍、汪庆2241；苏州穹窿山，凌萍萍850；吴县东山，凌萍萍613，凌萍萍、汪庆3345；无锡鼋头渚，凌汪2293；镇江北固山，凌萍萍、汪庆2702；句容茅山，凌萍萍、汪庆2327；连云港黄窝，凌萍萍508；连岛，凌萍萍、汪庆3051；柳河，凌萍萍303.

74. *Bryum lonchocaulon* C. Muell.世界广布成分

宜兴龙池，秦怀兰191.

75. *Bryum turbinatum* (Hedw.) Turn.

镇江焦山，凌萍萍、汪庆2693.

76. *Bryum tortifolium* Bird.北温带成分

宜兴龙头沟，李登科19576.

77. *Pohlia cruda* (Hedw.) Lindb. 世界广布成分

昆山亭林，凌萍萍、汪庆3271，3269，3270；吴县光福，凌萍萍715；无锡马山，凌萍萍996；连云港连岛，凌萍萍、汪庆3053；花果山，凌萍萍289；南通军山，凌萍萍、汪庆2913.

78. *Pohlia elongata* Hedw.

无锡鼋头渚，凌萍萍、汪庆2279，2293b;蠡园，吴鹏程、金岳杏1973；连云港保驾山，凌萍萍145；扬州瘦西湖，凌萍萍、汪庆2793；镇江南山，凌萍萍、汪庆2730；南通军山，凌萍萍、汪庆2915；狼山，凌萍萍、汪庆2925，2924.

79. *Rhodobryum roseum* (Hedw.) Limpr.北温带成分

宜兴乾元界，吴鹏程、金岳杏2119；老鹰界，吴鹏程、金岳杏2336，凌萍萍、汪庆2020；张公洞、吴鹏程、金岳杏2433；龙池，吴鹏程、金岳杏2292；吴县西山，吴鹏程、金岳杏1795，1833，1990；句容茅山，凌萍萍、汪庆2370，2367.

80. *Rhodobryum giganteum* (Schwaegr.) Par.

吴县西山，吴鹏程、金岳杏1846.

81. *Leptobryum pyriforme* (Hedw.) Wils.世界广布成分

南京中山植物园，凌萍萍912；昆山亭林，凌萍萍、汪庆3312，3323；吴县东山，凌萍萍、汪庆2517，2510；宜兴龙池，秦怀兰214.

Mniaceae提灯藓科

82. *Trachycystis microphylla* (Doz. et Molk.) Lindb.

吴县西山，吴鹏程、金岳杏1806，1824；光福，凌萍萍、汪庆2425，2430；东山白沙山，凌萍萍574；南通军山，凌萍萍、汪庆2878，2871；连云港，凌萍萍、汪庆3401,3412;无锡市植物园，凌萍萍971,972；鼋头渚,吴鹏程、金岳杏1935，凌萍萍、汪庆2299，2289；宜兴林场，凌萍萍、汪庆2009，2026，3357，2128；句容茅山，凌萍萍113，064；南京中山植物园，凌萍萍1574；紫金山下，凌萍萍928；常熟虞山，凌萍萍、汪庆2229.

83. *Plagiomnium cuspidatum* (Hedw.) T.Kop.

苏州天平山，凌萍萍、汪庆2595，凌萍萍795；西山，凌萍萍、汪庆2444，2443; 光福，凌萍萍、汪庆2405；扬州大明寺，凌萍萍、汪庆2807；无锡鼋头渚，凌萍萍、汪庆2281，2285b;宜兴林场，凌萍萍、汪庆2069；南通军山，凌萍萍、汪庆2883，2876；连云港花果山，凌萍萍、汪庆3225,3236；连岛，凌萍萍、汪庆3098；南京老山滴水珠，凌萍萍1640，1683；中山植物园，凌萍萍1586，1587；句容茅山，凌萍萍、汪庆2352，2359.

84. *Plagiomnium acutum* (Lindb.) T.kop.

句容茅山，凌萍萍065，071，114，116, 100；镇江南山,凌萍萍、汪庆2757；吴县东山，吴鹏程、金岳杏1734.

85. *Plagiomnium maximoviczii* (Lindb.) T.kop.
连云港后云台山，凌萍萍、汪庆2989，3116，3108；常熟虞山，李登科017638，017847；宜兴，邓懋彬7103，7104.

86. *Plagiomnium rostratum* (Schrad.) T.kop.ssp.vesicatum (Besch.) Nog.
宜兴龙池，李登科019649；句容宝华山，李登科020271.

87. *Plagiomnium rostratum* (Schrad.) T.kop.
连云港望海楼，凌萍萍126；句容茅山，凌萍萍092；宜兴左华，吴鹏程、金岳杏2043.

88. *Plagiomnium succulentum* (Mitt.) T.Kop.
宜兴龙池，吴鹏程、金岳杏2295.

89. *Mnium laevinerve* Card.
南京栖霞山，陈邦杰5986；连云港后，云台山，凌萍萍、汪庆2971a.

90. *Mnium marginatum* (Mitt.) P.Beauv.
连云港黄窝，凌萍萍、汪庆3029；大桅尖，凌萍萍、汪庆2989，2990.

91. *Mnium lycopodioides* Schwaegr.
昆山亭林，凌萍萍、汪庆3300.

92. *Mnium thomsonii* Schimp.
宜兴小黑沟，吴鹏程、金岳杏2370.

Rhizogoniaceae桧藓科

93. *Rhizogonium dozyanum* Lac. 中国-日本成分
家兴馨山，邓懋彬无号(标本室号436070)，吴鹏程、金岳杏2467，2453；宜兴深涧，秦怀兰52；省庄，张涛595，邓懋彬7094.

Bartramiaceae珠藓科

94. *Bartramia pomiformis* Hedw.北温带成分
宜兴省庄, 张涛644, 653, 秦怀兰348；老鹰界,吴鹏程、金岳杏2241，2171，2154.

95. *Philonotis thwaitesii* Mitt.中国-日本成分
句容茅山, 李登科020308；常熟虞山，李登科017829；宜兴乾元界，吴鹏程、金岳杏2087；善卷洞，吴鹏程、金岳杏2009，凌萍萍、汪庆2168，2158；吴县穹窿山，凌萍萍849；灵岩山，凌萍萍639，852；天平山，凌萍萍、汪庆2620；东山，凌萍萍579；连云港宿城，凌萍萍、汪庆3180，3204，3166，3132，3129；黄窝，凌萍萍、汪庆3037，3040，3048；大桅尖2974，2995；连岛，凌萍萍、汪庆3097，3091.

96. *Philonotis lancifolia* Mitt. 中国-日本成分
溧阳松岭，李登科18962；吴县灵岩山，刘仲苓32545；句容茅山，李登科020349; 常熟虞山, 李登科，17628；连云港宿城,凌萍萍、汪庆3195; 柳河，凌萍萍345；黄窝，凌萍萍、汪庆3019，万寿岩，凌萍萍187；吴县东山，凌萍萍644,839；南京中山植物园，凌萍萍934，943，930.

97. *Philonotis turneriana* (Schwaegr.) Mitt.
宜兴善卷洞，李登科019539；震泽洞庭西山，胡人亮113；常熟虞山，凌萍萍、汪庆2212.

98. *Philonotis hastata* (Dub.) Wijk. et Marg.
宜兴善卷洞，邓懋彬无号.

99. *Philonotis falcata* (Hook.) Witt.
连云港花果山，凌萍萍451, 315；黄窝，凌萍萍523；宿城，凌萍萍381，457.

Erpodiaceae树生藓科

100. *Aulacopilum japonicum* Broth. ex Card.中国–日本成分
溧阳横涧，李登科18845；常熟虞山，李登科17532，陈邦杰20936；宜兴省庄，张涛499.

101. *Venturiella sinensis* (Vent.) C.Muell. 东亚北美间断分布
产于长江以南平原近人处常见，生于枫杨及阔叶树干上.

Glyphomitriaceae高领藓科

102. *Glyphomitrium acuminatum* Broth.中国–日本成分
常熟虞山，李登科17567.

Orthotrichaceae 木灵藓科

103. *Orthotrichum courtoisii* Broth. et Par.中国特有成分
宜兴善卷洞，李登科00255；南京中山植物园，凌萍萍1651；宜兴老鹰界，凌萍萍、汪庆2076；苏州光福，凌萍萍、汪庆2391.

104. *Orthotrichum consobrinum* Card.中国 –日本成分
吴县西山，李登科18714.

105. *Drummondia sinensis* C.Muell.中国 –日本成分
吴县石公山，李登科18714.

106. *Macromitrium gymnostomum* Sull. et lesq.中国 –日本成分
宜兴湖汶，吴鹏程、金岳杏2461；吴县西山，胡人亮12.

107. *Macromitrium japonicum* Doz.et Molk.中国 –日本成分
连云港花果山，凌萍萍256，192；锦屏山，邓懋彬无号(标本号496969).

108. *Macromitrium courtoisii* Broth. et Par.中国特有成分
吴县东山，吴鹏程、金岳杏1880；西山，1826；灵岩山，凌萍萍820；宜兴馨山，吴鹏程、金岳杏2457；南通狼山，凌萍萍、汪庆2927；连云港黄窝，凌萍萍、汪庆3050；柳河，凌萍萍327，341.

109. *Macromitrium ousiense* Broth. et Par.中国特有成分
模式标本采于无锡，未见该标本.

110. *Macromitrium ferriei* Card. et Ther.中国 –日本成分

宜兴老鹰咀，李登科19585，凌萍萍、汪庆2008；太华山，吴鹏程、金岳杏2078；龙池，吴鹏程、金岳杏2305，2290；馨山，吴鹏程、金岳杏2442；省庄，秦怀兰296，323；茗岭，秦怀兰134，135，张涛447；连云港柳河，凌萍萍201；吴县东山，胡人亮8.

111.*Schlotheimia pungens* Bartr. 中国-日本成分— S. charrieri Ther. et P.de La Varde 宜兴省庄，张涛593.

Racopilaceae卷柏藓科

112.*Racopilum aristatum* Mitt.中国-日本成分

宜兴龙池，高彩华26937，李登科19645；乾元界，吴鹏程、金岳杏2053，2083，2107；张公洞，吴鹏程、金岳杏2426，2435.

Hedwigiaceae 虎尾藓科

113.*Hedwigia ciliata* (Hedw.) Ehrh. ex P.Beauv.世界广布成分

产于本区各低山区.

Trachypodaceae扭叶藓科

114.*Trachypus humilis* Lindb.热带亚洲一大洋洲成分

宜兴龙池，吴鹏程、金岳杏2269；乾元界，吴鹏程、金岳杏2131，2133，2130.

Myuriaceae金毛藓科

115.*Oedicladium rufescens* (Reinw. et Hornsch.) Mitt.

宜兴老鹰界，吴鹏程、金岳杏2177.

Meteoriaceae蔓藓科

116.*Meteorium buchananii* (Bird.) Broth.

常熟虞山，李登科17551；吴县灵岩寺，李登科18839；句容宝华山，李登科20254；宜兴小黑沟，吴鹏程、金岳杏2367，2340，凌萍萍、汪庆3392；吴县邓尉山，陈邦杰2211.

117.*Aerobryopsis walichii* (Brid.) Fleisch.

宜兴馨山，吴鹏程、金岳杏2444.

118.*Pseudobarbella attenuata* (Thwait. et Mitt.) Nog.

宜兴龙池，李登科19658；老鹰咀，李登科19604.

119.*Pseudobarbella laxifolia* Nog.

宜兴善卷洞，吴鹏程、金岳杏2021；茗岭，秦怀兰22，119.

120.*Meteorium subpolytrichum* (Besch.) Broth.中国-日本成分

溧阳平桥，李登科18904；宜兴龙池，吴鹏程、金岳杏2321.

121.*Aerobryopsis subdivergens* (Broth.) Broth.

宜兴龙池，吴鹏程、金岳杏2256.

122.*Chrysocladium retrorsum* (Mitt.) Fleisch.

宜兴龙池，吴鹏程、金岳杏2267；老鹰界，吴鹏程、金岳杏2181；宜兴，陈封怀13636；省庄，张涛.

Neckeraceae平藓科

123.*Neckera yezoana* Besch.中国-日本成分

南京梅花山，李登科16603.

124.*Homaliodendron exiguum* (Bosch. et Lac.) Fleisch.旧世界热带分布

宜兴龙池，吴鹏程、金岳杏2264.

125.*Homalia trichomanoides* (Hedw.) B.S.G.

宜兴小黑沟，李登科020405；馨山，吴鹏程、金岳杏3471.

126.*Porotrichum perplexans* Dix.

《中国藓类植物属志》

127.*Thamnobryum sandei* (Besch.) Iwats.

宜兴乾元界，吴鹏程、金岳杏2095，2099，2060，2138，2030；老鹰界，吴鹏程、金岳杏2169, 2137;小黑沟，吴鹏程、金岳杏2369，2341，2381，凌萍萍、汪庆3372; 龙池, 秦怀兰205；溧阳横涧，李登科18869；连云港大桅尖,凌萍萍、汪庆3099.

Hookeriaceae 油藓科

128.*Distichopyllum maibarae* Besch.热带亚洲及太平洋成分

宜兴小黑沟，吴鹏程、金岳杏2383，2388；老鹰界，吴鹏程、金岳杏2216；太华，吴鹏程、金岳杏2219；馨山，吴鹏程、金岳杏2469，2440.

129.*Hookeria acutifolia* Hook. et Grev.泛热带成分

宜兴老鹰界，吴鹏程、金岳杏2197，2232；宜兴龙池，凌萍萍、汪庆2211；连云港柳河，凌萍萍、汪庆3156；宿城，凌萍萍、汪庆31916.

Fabroniaceae 碎米藓科

130.*Juratzkea sinensis* Fleisch. ex Broth. 东亚北美间断成分

吴县茅蓬，李登科19472；南京中山植物园，凌萍萍、汪庆3436.

131.*Schwetschkea matsumurae* Besch.中国-日本成分

苏州虎丘，凌萍萍789.

132.*Schwetschkea courtoisii* Broth.et Par.中国特有成分

模式产地为上海徐家汇；苏州留园，凌萍萍、汪庆2568，2564；拙政园，凌萍萍汪庆2555；吴县天平山，凌萍萍、汪庆2631；苏州狮子林，凌萍萍、汪庆2539；虎丘，凌萍萍、汪庆2580，2588；吴县东山，凌萍萍741；灵岩山，凌萍萍664，793；扬州瘦西湖，凌萍萍、汪庆2780-b,2772；南京宝华山，凌萍萍1559；句容茅山，凌萍萍082.

133.*Schwetschkea sinica* Broth. et Par.中国特有成分

南京为模式产地；南京中山植物园，凌萍萍、汪庆3438；苏州虎丘，凌萍萍789，751，853；拙政园，凌萍萍860；吴县光福，凌萍萍848.

134.*Fabronia matsumurae* Besch.

宜兴善卷洞，李登科268；句容茅山，凌萍萍、汪庆2068；南京栖霞山，秦怀兰408；无锡，邓懋彬无号.

135.*Schwetschkeopsis fabronia* (Schwaegr.) Broth.东亚北美间断成分

南京灵谷寺，李登科16579；宜兴小黑沟，吴鹏程、金岳杏2315，2363，2361；乾元界，吴鹏程、金岳杏2064，2089，2076；龙池，吴鹏程、金岳杏2304；老鹰界，吴鹏程、金岳杏2202；省庄，张涛616；吴县西山，吴鹏程、金岳杏1854，17946；连云港柳河，凌萍萍331；黄窝，凌萍萍、汪庆3025；宿城，凌萍萍414；溧阳横涧，李登科18870；平桥，李登科18911.

Leskeaceae薄罗藓科

136.*Regmatodon declinatus* (Hook.) Brid.

137.*Lindbergia sinensis* (C.Muell.) Broth.

吴县虎丘，李登科6907；南京，复旦145；常熟虞山，李登科17618.

138.*Lindbergia brachyptera* (Mitt.) Kindb.

南京灵谷寺，李登科16572；苏州大公园，凌萍萍、汪庆2525；南京明孝陵，凌萍萍1601；中山植物园，凌萍萍1676，凌萍萍、汪庆3439，3437.

139.*Lindbergia japonica* Card.中国–日本成分

南京灵谷寺，李登科16607；吴县西山，胡人亮112，李登科6807，凌萍萍、汪庆2458；南京中山植物园，凌萍萍、汪庆3440.

140.*Pseudoleskeopsis zippelii* (Doz. et Molk.) Broth.中国–日本成分

宜兴老鹰岩，吴鹏程、金岳杏2213，吴鹏程、金岳杏2162；南京栖霞山，秦怀兰403；连云港万寿岩，凌萍萍198；句容茅山，凌萍萍106.

141.*Pseudoleskeopsis decurvata* (Mitt.) Broth.

常熟虞山，李登科17809，17534；宜兴茗岭，张涛440.

142.*Leskeella nervosa* (Brid.) Loesk.

苏州虎丘，凌萍萍、汪庆2587，凌萍萍768；吴县天平山，凌萍萍578；苏州拙政园，凌萍萍、汪庆2554；吴县灵岩山，吴鹏程、金岳杏1909；宜兴张公洞，吴鹏程、金岳杏2420.

143.*Leskea polycarpa* Ehrh. ex Hedw.

吴县东山，吴鹏程、金岳杏1634；宜兴老鹰界，吴鹏程、金岳杏2193.

144.*Lescuraea incurvata* (Hedw.) Iawt.

溧阳横涧，李登科18894；苏州虎丘，凌萍萍822；吴县西山，凌萍萍、汪庆2458；连云港柳河，凌萍萍303.

Thuidiaceae羽藓科

145.*Haplohymenium triste* (Ces.) Kindb.北温带成分

溧阳横涧，李登科18843；常熟虞山，李登科17472；宜兴龙池，吴鹏程、金岳杏2309；吴县西山，吴鹏程、金岳杏1848，1874，吴鹏程、金岳杏1730，凌萍萍、汪庆2498；无锡蠡园，吴鹏程、金岳杏1967，1966，1972；东山，李登科19380，吴鹏程、金岳杏1730，凌萍萍、汪庆2498；灵岩寺，李登科19370；南京紫金山，李登科16609.

146.*Anomodon rotundatus* Par. et Broth.中国特有成分

吴县东山，凌萍萍622，吴鹏程、金岳杏1647.

147.*Anomodon rugelii* (C.Muell.) Keissl.北温带成分

昆山，李登科04120；句容茅山，李登科020302，凌萍萍、汪庆2365；宜兴张公洞，吴鹏程、金岳杏2428，2434；善卷洞，吴鹏程、金岳杏2019，1999，1997；南京栖霞山，凌萍萍033.

148.*Anomodon viticulosus* (Hedw.) Hook.et Tayl.北温带成分

句容茅山，李登科20285，凌萍萍098；宜兴善卷洞，李登科00214；龙池,吴鹏程金岳杏2308; 小黑沟, 吴鹏程、金岳杏2349，2350；老鹰界，凌萍萍、汪庆2048，吴鹏程、金岳杏2214；南京灵谷寺，李登科16584；震泽西山，胡人亮100.

149.*Anomodon giraldii* C.Muell.中国 –日本成分

溧阳横涧，李登科18875；宜兴龙池，吴鹏程、金岳杏2320；小黑沟，吴鹏程、金岳杏2346；乾元界, 吴鹏程、金岳杏2025；南京吸孝陵, 凌萍萍1623，1624；栖霞山，秦怀兰400.

150.*Anomodon minor* ssp. *integerrimus* (Mitt.) Iwats.

南京灵谷寺，李登科16574；溧阳横涧，李登科18937.

151.*Herpetineuron toccoae* (Sull. et Lesq.) Card. 泛热带成分

昆山亭林，凌萍萍、汪庆3342; 常熟虞山，凌萍萍、汪庆2251; 宜兴善卷洞，凌萍萍、汪庆2131；老鹰界，吴鹏程、金岳杏2134；小黑沟，吴鹏程、金岳杏2405；吴县灵岩山，吴鹏程、金岳杏1899，1890，凌萍萍728, 585；东山，吴鹏程、金岳杏1626，凌萍萍677，629；西山, 凌萍萍、汪庆2471; 穹窿山，凌萍萍704, 光福，凌萍萍713，凌萍萍、汪庆2403，2102，2421，2427；无锡鼋头渚，吴鹏程、金岳杏1961，1946；镇江焦山，凌萍萍、汪庆2670，2669; 南通军山, 凌萍萍、汪庆2867, 2865；南京中山植物园，凌萍萍1697；连云港大桅尖，凌萍萍、汪庆3004；柳河，凌萍萍228，322；万寿岩，凌萍萍201；花果山，凌萍萍286；望海楼，凌萍萍120，146.

152.*Claopodium prionophyllum* Broth.

宜兴龙池，吴鹏程、金岳杏2332，2282；扬州瘦西湖，凌萍萍、汪庆2776；吴县西山，吴鹏程、金岳杏1858；苏州大公园，凌萍萍、汪庆2521；南京中山植物园，凌萍萍1669.

153.*Claopodium assurgens* (Sull. et Lesg.) Card.

宜兴老鹰界，凌萍萍、汪庆2078，2080，2067；无锡鼋头渚，凌萍萍、汪庆2285，2282；连云港连岛，凌萍萍、汪庆3071.

154.*Claopodium aciculum* (Broth.) Broth.中国－日本成分

常熟，李登科17604；震泽西山，胡人亮107；南京，李登科16566；宝华山，凌萍萍、汪庆1525，1555；中山植物园，凌萍萍939；宜兴乾元界，吴鹏程、金岳杏2137，2507；龙池，吴鹏程、金岳杏2291；连云港大桅尖，凌萍萍、汪庆3206，3395；连岛，凌萍萍、汪庆3069；花果山，凌萍萍317；吴县东山，凌萍萍、汪庆2507；西山，吴鹏程、金岳杏1860.

155.*Bryohaplocladium microphyllum* (Hedw.) Watan. et Iwats.

产于本地区平原及山地.

156.*Bryohaplocladium angustifolium* (Hamp. et C.Muell.) Watan.et Iwats.

泛热带分布.

157.*Thuidium glaucinum* (Mitt.) Bosch. et Lac.

宜兴小黑沟，李登科20371，凌萍萍、汪庆3370；老鹰界，凌萍萍、汪庆2065；吴县西山，吴鹏程、金岳杏1877.

158.*Thuidium kanedae* Sak.中国－日本成分

常熟虞山，李登科017781；溧阳横涧，李登科18883；宜兴小黑沟，吴鹏程、金岳杏2411；吴县西山，吴鹏程、金岳杏1779；昆山，凌萍萍、汪庆 3340.

159.*Thuidium cymbifolium* (Doz.et Molk.) Doz. et Molk. 旧世界热带成分

宜兴小黑沟，吴鹏程、金岳杏2352，2360，凌萍萍、汪庆3373，3387；龙池，吴鹏程、金岳杏2288；乾元界，吴鹏程、金岳杏2045，2090；老鹰界，吴鹏程、金岳杏2294，2210，凌萍萍、汪庆2042；馨山，吴鹏程、金岳杏2481；吴县西山，吴鹏程、金岳杏1804，1781；东山，吴鹏程、金岳杏1726，1694，1658；南通狼山，凌萍萍、汪庆2932；无锡鼋头渚，吴鹏程、金岳杏1954，1927.

Amblystegiaceae 柳叶藓科

160.*Campylium chyrysophyllum* (Brid.) J.Lange

南京紫金山，凌萍萍1591；常熟虞山，凌萍萍、汪庆2234；连云港大桅尖，凌萍萍、汪庆2991，2987；黄窝L.W.3007，凌萍萍515.

161.*Campylium stellatum* (Hedw.) C.Jens.

连云港大桅尖，凌萍萍529.

162.*Amblystegium riparium* (Hedw.) B.S.G.北温带分布

常熟虞山，李登科017465；宜兴玉女潭，高彩华26824；南京中山植物园，凌萍萍895；徐州铜山，凌萍萍、汪庆2958，2960.

163.*Cratoneuron filicinum* (Hedw.) Spruce

宜兴老鹰界，凌萍萍、汪庆2190.

164.*Amblystegium jungermannioides* (Brid.) Smith.东亚－北美间断成分

宜兴省庄，张涛无号.

165.*Hygrohypnum eugyrium* (B.S.G.) Broth.北温带成分

宜兴老鹰咀，高彩华26744，吴鹏程、金岳杏2160，2178，2183，凌萍萍、汪庆2046；小黑沟，高彩华26886，吴鹏程、金岳杏2408；馨山,吴鹏程、金岳杏2468,2487；宜兴，邓懋彬7099，凌萍萍、汪庆3358，3389；句容茅山，凌萍萍080.

166.*Hygrohypnum luridum* (Hedw.) Jenn.

宜兴茗岭，张涛434，429；吴县西山，胡人亮122.

Brachytheciaceae青藓科

167.*Homalothecium laevisetum* Lac.中国－日本成分

溧阳平桥，李登科18906，18933；昆山亭林，凌萍萍、汪庆3295，3268；宜兴龙池，吴鹏程、金岳杏2319，2253，2339；乾元界，吴鹏程、金岳杏2091；句容茅山，凌萍萍、汪庆2366；南京宝华山，凌萍萍1563.

168.*Palamocladium macrostegium* (Sull. et Lesq.) Iwats.et Tak.中国－日本成分

昆山亭林，凌萍萍、汪庆3292，李登科4124；宜兴小黑沟，凌萍萍、汪庆3362，3364；乾元界，吴鹏程、金岳杏2118；省庄，张涛639.

169.*Palamocladium euchloron* (C.Muell.) Wijk. et Marg.

昆山亭林，凌萍萍、汪庆3341.

170.*Brachythecium pinnatum* Tak. 中国－日本成分

宜兴乾元界，吴鹏程、金岳杏2133；吴县天平山，凌萍萍、汪庆2601；东山，凌萍萍8311；南京宝华山，凌萍萍1513；句容茅山，凌萍萍59；连云港花果山，凌萍萍200，299；大桅尖，凌萍萍、汪庆3000.

171.*Brachythecium wichurae* (Broth.) Par. 中国－日本成分

宜兴小黑沟，李登科20435；善卷洞，李登科242.

172.*Brachythecium nakazimae* Ihs.中国－日本成分

南京中山植物园，凌萍萍1661；宜兴老鹰界，凌萍萍、汪庆2109，2110；吴县东山，吴鹏程、金岳杏1682，1624；镇江南山，凌萍萍、汪庆2736，2716，2744，2746；扬州瘦西湖，凌萍萍、汪庆2301；昆山亭林，凌萍萍、汪庆3259，3265，3334，3332.

173.*Brachythecium campestre* (C.Muell.) B.S.G.北温带成分

南京玄武湖，鲁玲6036；中山植物园，凌萍萍1599，1675，1620；燕子矶，秦怀兰41，39，29；明孝陵，凌萍萍1623；栖霞山，凌萍萍049；宜兴深涧，秦怀兰57;善卷洞，秦怀兰268；吴县东山，凌萍萍、汪庆2467；西山，凌萍萍、汪庆2467，2445，2443；光福，凌萍萍、汪庆2439，2438，2436；苏州留园，凌萍萍、汪庆2574.

174.*Brachythecium buchananii* (Hook.) Jaeg.中国－日本成分

宜兴善卷洞，李登科222，吴鹏程、金岳杏1995；南京宝华山，凌萍萍1546；连云港柳河，凌萍萍225；大桅尖，凌萍萍、汪庆2985；宿城，凌萍萍、汪庆3187；吴县东山，凌萍萍865；溧阳横涧，李登科18863；句容茅山，凌萍萍、汪庆2342.

175.*Brachythecium reflexum* (Starke) B.S.G.北温带成分

天县西山，李登科18791，凌萍萍、汪庆2449；天平山，凌萍萍、汪庆2632, 2618；灵岩山，凌萍萍794；宜兴老鹰界，凌萍萍、汪庆3356，3382，3381，吴鹏程、金岳杏2246，2227；善卷洞，秦怀兰251，凌萍萍、汪庆2157；乾元界，吴鹏程、金岳杏2049；镇江，凌萍萍、汪庆2695；苏州虎丘，凌萍萍、汪庆3577，凌萍萍791；连云港大桅尖，凌萍萍、汪庆2988，3407；黄窝，凌萍萍、汪庆3028；常熟虞山，凌萍萍、汪庆2255.

176.*Brachythecium brotheri* Par.中国 –日本成分

吴县光福，凌萍萍、汪庆2434；西山，凌萍萍、汪庆2463；东山，凌萍萍、汪庆2488；天平山，凌萍萍、汪庆2597,2626；苏州拙政园,凌萍萍、汪庆2550，2552；苏州大学，凌萍萍、汪庆2527，2528；苏州大公园，凌萍萍、汪庆2520；宜兴老鹰界，凌萍萍、汪庆3363；句容茅山，凌萍萍、汪庆3217；昆山亭林3286.

177.*Brachythecium velutinum* (Hedw.) B.S.G.北温带成分

南京太平门，李兆兰106084；宝华山，凌萍萍1560，1534，1571，1554；宜兴乾元界，吴鹏程、金岳杏2134；善卷洞，吴鹏程、金岳杏2010；吴县东山，吴鹏程、金岳杏1655，1618；无锡蠡园，吴鹏程、金岳杏1980，1984；昆山亭林，凌萍萍、汪庆3272；镇江焦山，凌萍萍、汪庆2659；南山，凌萍萍、汪庆2734；句容茅山凌萍萍、汪庆2330；南通军山，凌萍萍、汪庆2898，2907，2899；扬州瘦西湖，凌萍萍、汪庆2778；个园，凌萍萍、汪庆2848；大明寺，凌萍萍、汪庆2803-b；连云港花果山，凌萍萍、汪庆3214，3216.

178.*Brachythecium populeum* (Hedw.) B.S.G.北温带成分

吴县西山，李登科6833；宜兴老鹰界，吴鹏程、金岳杏2252.

179.*Brachythecium plumosum* (Hedw.) B.S.G.北温带成分

宜兴善卷洞，李登科224.

180.*Brachythecium glauculum* C.Muell.

句容茅山，凌萍萍、汪庆2363，2337.

181.*Bryhnia novae-angliae* (Sull.et Lesq.) Grout 北温带成分

宜兴龙池，李登科19655.

182.*Bryhnia sublaevifolia* Broth.et Par.中国 –日本成分

宜兴龙头沟，李登科19574.

183.*Myuroclada maximowiczii* (Borszcz.) Steere et Schof.

北温带成分，产于本地区平原或山地.

184.*Rhynchostegium contractum* Card.中国 –日本成分

宜兴龙头沟，李登科19557；老鹰界，吴鹏程、金岳杏2221；张公洞，吴鹏程、金岳杏2439；昆山亭林，凌萍萍、汪庆3302；吴县西山，凌萍萍、汪庆2485；扬州大明寺，凌萍萍、汪庆2810；镇江焦山，凌萍萍、汪庆2689；南京燕子矶，秦怀兰40.

185.*Rhynchostegium plumosum* Ther.中国 -日本成分

宜兴老鹰界，李登科19620，凌萍萍、汪庆2062.

186.*Rhynchostegium pallidifolium* (Mitt.) Jaeg.中国 -日本成分

苏州虎丘，凌萍萍766；吴县光福，凌萍萍799；宜兴小黑沟，凌萍萍、汪庆2136，2137-b；老鹰界，凌萍萍、汪庆2052-a,2053,2049；龙池，吴鹏程、金岳杏2307,2258, 邓懋彬7084；南京紫金山，凌萍萍879，03；宝华山，凌萍萍1551，1536.

187.*Oxyrrhychium laxirete* Broth. et Card. 中国-日本成分

宜兴老鹰界，李登科19590，吴鹏程、金岳杏2195;善卷洞,吴鹏程、金岳杏2014；张公洞，吴鹏程、金岳杏2427；小黑沟，吴鹏程、金岳杏2386；昆山亭林，凌萍萍、汪庆3329，3330；吴县东山，吴鹏程、金岳杏1714，1693，1723；西山，凌萍萍、汪庆2456；苏州留园，凌萍萍、汪庆2556；扬州何园，凌萍萍、汪庆2857，2855；瘦西湖，凌萍萍、汪庆2770；大明寺，凌萍萍2803；连云港黄窝，凌萍萍、汪庆3022，3018; 连岛, 凌萍萍、汪庆3081；大桅尖，凌萍萍、汪庆3101；柳河，凌萍萍、汪庆3121；万寿岩，凌萍萍182, 151.

188.*Eurhynchium savatieri* Schimp. ex Besch.中国 -日本成分

连云港宿城，凌萍萍、汪庆3177；柳河，凌萍萍329；万寿岩，凌萍萍195；吴县东山，凌萍萍548，689；宜兴，邓懋彬7096，7107.

189.*Eurhynchium riparioides* (Hedw.) Richard.世界广布成分

连云港黄窝，凌萍萍、汪庆3032，3024，3014,3033；大桅尖,凌萍萍、汪庆3104，3105；柳河，凌萍萍、汪庆3154，3150，3158，3155；宿城，凌萍萍138；苏州留园，凌萍萍、汪庆2537.

190.*Eurhynchium pulchellum* Jenn.世界广布成分

宜兴龙池，李登科19660；馨山，吴鹏程、金岳杏2464，2477；善卷洞，秦怀兰261，吴鹏程、金岳杏2003；老鹰界，吴鹏程、金岳杏2208；连云港玉女峰，凌萍萍261；花果山，凌萍萍、汪庆3233；大桅尖，凌萍萍、汪庆2930.

191.*Eurhynchium praelongum* (Hedw.) B.S.G.

宜兴张公洞，吴鹏程、金岳杏2419；省庄，张涛494；连云港宿城, 凌萍萍、汪庆3191.

Entodontaceae绢藓科

192.*Entodon viridulus* Card.中国 -日本成分

宜兴老鹰岩，凌萍萍、汪庆2041，2054，2064,2035；吴鹏程、金岳杏2201，2204；小黑沟，凌萍萍、汪庆2148, 2142；乾元界，吴鹏程、金岳杏2114；南京宝华山，凌萍萍1535，1562.

193.*Entodon micropodus* (Hedw.) C.Muell.北温带成分(东亚一北美间断成分)

宜兴老鹰岩，凌萍萍、汪庆2028，2033，2650；小黑沟，吴鹏程、金岳杏

2374，2365；乾元界，吴鹏程、金岳杏2063a;无锡鼋头渚，吴鹏程、金岳杏1960；吴县西山，吴鹏程、金岳杏1828；昆山亭林，凌萍萍、汪庆3334；溧阳平桥，李登科18925，18928.

194.*Entodon compressus* (Hedw.) C.Muell.北温带成分

宜兴林场，凌萍萍、汪庆2119，2114，2111；善卷洞，凌萍萍、汪庆2173，2154，2579；吴县光福，凌萍萍、汪庆748，2432，2460，2389；天平山，凌萍萍、汪庆2603；西山，凌萍萍、汪庆2389，2462；南京中山植物园，凌萍萍1618，1611；紫金山，凌萍萍1592，1596；栖霞山，凌萍萍045，046；无锡鼋头渚，凌萍萍、汪庆2292.

195.*Entodon luridus* (Griff.) Jaeg.中国 –日本成分

宜兴小黑沟，吴鹏程、金岳杏2407，2394；老鹰岩，吴鹏程、金岳杏2158；茗岭，吴鹏程、金岳杏2301；溧阳平桥，李登科1194; 吴县东山, 吴鹏程,金岳杏1697；灵岩山，吴鹏程、金岳杏1894，凌萍萍562；扬州瘦西湖，凌萍萍、汪庆2767，2768，2769；苏州光福，凌萍萍、汪庆2419，2382，2412；虎丘，凌萍萍、汪庆2576；平山，凌萍萍、汪庆2616；句容茅山，凌萍萍104; 连云港柳河，凌萍萍、汪庆3174；大桅尖，凌萍萍、汪庆2988，2976；连岛，凌萍萍、汪庆3132；花果山，凌萍萍、汪庆3229.

196.*Entodon cladorrhizans* (Hedw.) C.Muell.北温带成分

宜兴，邓懋彬7099-c；吴县光福，凌萍萍783；句容茅山，凌萍萍102；镇江焦山，凌萍萍、汪庆2662，2667；南山，凌萍萍、汪庆2375；连云港大桅尖，凌萍萍、汪庆3107；宿城，凌萍萍、汪庆3185，3186；常熟虞山，凌萍萍、汪庆2246；南京宝华山，凌萍萍1521，1532，1559，1526.

197.*Entodon sullivantii* (C.Muell.) Lindb.

吴县西洞庭山，胡人亮34，103；常熟虞山，凌萍萍、汪庆2230，李登科17844；常州，陈邦杰35；镇江金山，凌萍萍、汪庆2714；吴县光福，凌萍萍、汪庆2407；东山，凌萍萍652，吴鹏程、金岳杏1657；宜兴深涧，秦怀兰12；无锡鼋头渚，吴鹏程、金岳杏1941，1934；马山，凌萍萍1003；连云港宿城，凌萍萍、汪庆3183.

Plagiotheciaceae棉藓科

198.*Plagiothecium neckeroideum* B.S.G.

宜兴老鹰咀，李登科19621，凌萍萍、汪庆2061，邓懋彬7086；连去港，凌萍萍、汪庆3396.

199.*Plagiothecium cavifolium* (Brid.) Iwats.北温带成分

宜兴老鹰咀，李登科19802；乾元界，吴鹏程、金岳杏2040，2061,2110；老鹰界，吴鹏程、金岳杏2240，2243；龙池，吴鹏程、金岳杏2328，2335；苏州虎丘，凌萍萍752；连云港连岛，凌萍萍、汪庆3095; 大桅尖, 凌萍萍、汪庆3002,2971-b；吴县西山，吴鹏程、金岳杏1807.

200.*Plagiothecium nemorale* (Mitt.) Jaeg.温带亚洲成分

宜兴龙池，李登科19653；吴县灵岩山，吴鹏程、金岳杏1883，凌萍萍596；东山吴鹏程、金岳杏1706，凌萍萍669，746，759；光福，凌萍萍、汪庆2429，2401；西山，凌萍萍、汪庆2468；天平山，凌萍萍、汪庆3625；句容茅山，凌萍萍、汪庆2341，凌萍萍094；苏州虎丘，凌萍萍、汪庆2589；昆山亭林，凌萍萍、汪庆3219；南通军山，凌萍萍、汪庆2889，2877，2869.

201.*Plagiothecium platyphyllum* Moenk.温带亚洲成分

宜兴老鹰咀，李登科19631；句容茅山，凌萍萍、汪庆2340; 吴县, 邓尉山、陈邦杰2213；无锡马山，凌萍萍1004；镇江焦山，凌萍萍、汪庆2371.

202.*Plagiothecium sylvaticum* (Brid.) B.S.G.

宜兴老鹰界，吴鹏程、金岳杏2209；李登科19614.

203.*Plagiothecium laevigatum* Besch.

宜兴老鹰界，吴鹏程、金岳杏2220；小黑沟, 吴鹏程、金岳杏2355; 龙池，吴鹏程、金岳杏2376，2266；乾元界，吴鹏程、金岳杏2059，2094，2120，2053；吴县东山，凌萍萍691，787.

Sematophyllaceae锦藓科

204.*Brotherella recurvans* (Michx.) Fleisch.

宜兴小黑沟，李登科20399；龙池，凌萍萍、汪庆2205，2207；老鹰岩，凌萍萍、汪庆2011.

205.*Brotherella erythrocaulis* (Mitt.) Fleisch.

宜兴乾元界，吴鹏程、金岳杏2194,2063,2050；吴县西山，吴鹏程、金岳杏1743.

206.*Brotherella falcatula* Broth.

宜兴老鹰岩，吴鹏程、金岳杏2153，2230，2234，2174，2185，2203.

207.*Brotherella henonii* (Dub.) Fleisch.

句容茅山，李登科20324.

208.*Sematophyllum pulchellum* (Card.) Broth.中国一日本成分

南京灵谷寺，李登科16589；紫金山，凌萍萍16；宜兴茗岭，秦怀兰92；老鹰岩，吴鹏程、金岳杏2182；溧阳平桥，李登科18930.

Hypnaceae灰藓科

209.*Gollania turgens* (C.Muell.) Ando

宜兴老鹰界，吴鹏程、金岳杏2147；龙池，秦怀兰153；省庄，秦怀兰307；茗岭，秦怀兰106.

210.*Homomallium connexum* (Card.) Broth.中国 -日本成分

宜兴乾元界，吴鹏程、金岳杏2128，2121，2079，2070，2092；龙池,秦怀兰191，吴鹏程、金岳杏2314，吴县光福，凌萍萍828，626，凌萍萍、汪庆2397，2416，2424；茅蓬，李登科19500；西山，吴鹏程、金岳杏1766;常熟虞山，李登科17489；苏州虎丘，凌萍萍754.

211.*Homomallium japonico-adnatum* (Broth.) Broth.
吴县西山，吴鹏程、金岳杏1766.
212.*Hypnum plumaeforme* Wils.温带亚洲成分
昆山亭林，凌萍萍、汪庆3304；常熟虞山，凌萍萍、汪庆2219，2217；宜兴，邓懋彬7098；吴县光福，凌萍萍、汪庆2376；天平山，凌萍萍、汪庆2622；无锡马山，凌萍萍997，1019，956；鼋头渚，吴鹏程、金岳杏1920；句容茅山，凌萍萍汪庆2315，2318；南通军山，凌萍萍、汪庆2868；南京中山陵，凌萍萍1576；中山植物园，凌萍萍1613；连云港大桅尖，凌萍萍、汪庆2972；花果山，凌萍萍、汪庆3249，3226；连岛，凌萍萍、汪庆3059，3061，3092；黄窝3006.
213.*Hypnum erectiusculum* Sull. et Lesq.中国－日本成分
宜兴龙池，李登科19669，吴鹏程、金岳杏2268，凌萍萍、汪庆2202.
214.*Hypnum callichroum* Brid.北温带成分
吴县东山，李登科6751，凌萍萍749；光福，凌萍萍718；穹窿山，凌萍萍694，666，657；灵岩山，凌萍萍608；宜兴老鹰界，吴鹏程、金岳杏2412；罄山，吴鹏程、金岳杏2480，2483；乾元界，吴鹏程、金岳杏2031；昆山亭林，凌萍萍、汪庆3318；南通军山，凌萍萍、汪庆2879；连云港柳河，凌萍萍328，凌萍萍、汪庆3153；大桅尖，凌萍萍、汪庆3102；连岛，凌萍萍、汪庆3353.
215.*Hypnum hamulosum* B.S.G.北温带成分
常熟，李登科017509；宜兴林场，凌萍萍、汪庆2117；龙池，高彩华26933；昆山亭林，凌萍萍、汪庆3303，3313；吴县西山，吴鹏程、金岳杏1842；东山，凌萍萍660，775.
216.*Ectropothecium zollingeri* (C.Muell.) Jaeg.中国－日本成分
宜兴老鹰界，凌萍萍、汪庆2209；罄山，邓懋彬无号；吴县光福，凌萍萍、汪庆2390；镇江焦山，凌萍萍、汪庆2663，2674；扬州个园，凌萍萍、汪庆2831；南京栖霞山，凌萍萍037，044；宝华山，凌萍萍154，1523；连云港，凌萍萍、汪庆3400；大桅尖，凌萍萍、汪庆982；宿城，凌萍萍435.
217.*Isopterygium pohliaecarpum* (Sull. et Lesq.) Jaeg.中国－日本成分
溧阳，李登科18952；宜兴善卷洞，李登科18538；龙池，凌萍萍、汪庆2179，2204；张涛32，吴鹏程、金岳杏2306；湖汶，邓懋彬7084；无锡马山，凌萍萍975；鼋头渚，凌萍萍、汪庆2286；吴县东山，吴鹏程、金岳杏1718；连云港花果山，凌萍萍、汪庆3228.
218.*Isopterygium courtoisii* Broth. et Par.中国特有成分
宜兴善卷洞，吴鹏程、金岳杏2004；乾元界，吴鹏程、金岳杏2080，2093；连云港柳河，凌萍萍236.
219.*Isopterygium minutirameum* (Schimp.) Jaeg.
常熟虞山，凌萍萍、汪庆2240，2231；宜兴林场，凌萍萍、汪庆3384；吴县西山，凌萍萍、汪庆2474，李登科18736；东山，凌萍萍698；无锡鼋头渚，凌萍萍、汪庆2302；马山，凌萍萍 982.

220.*Taxiphyllum taxirameum* (Mitt.) Fleisch.中国 -日本成分

地区各地平原或山地；常熟虞山，凌萍萍、汪庆2252；昆山亭林，凌萍萍、汪庆3298；兴善卷洞，凌萍萍、汪庆2162，吴鹏程、金岳杏2012，1994，1996；吴县东山，吴鹏程、岳杏1672，1716，1673，凌萍萍、汪庆2508；灵岩山，凌萍萍、汪庆2636，吴鹏程、金岳杏1891；天平山，凌萍萍、汪庆2604，2599；无锡鼋头渚，凌萍萍、汪庆2280；镇江焦山，凌萍萍、汪庆2657，2661；南山，凌萍萍、汪庆2721，2730；句容茅山，凌萍萍、汪庆2312，33；南通军山，凌萍萍、汪庆2905；狼山，凌萍萍、汪庆2928；扬州瘦西湖，凌萍萍、汪庆2785；南京燕子矶，秦怀兰360；中山植物园，凌萍萍1614；连云港墟沟，凌萍萍480；宿城，凌萍萍169；连岛，凌萍萍3062；花果山，凌萍萍、汪庆3224.

221.*Taxiphyllum subarcuatum* (Broth.) Iwats. 中国一日本成分

常熟虞山，李登科17530；昆山亭林，凌萍萍、汪庆3283；南京宝华山，凌萍萍1552，1553；金山，凌萍萍015;宜兴龙池，凌萍萍、汪庆2186，2191，2192；老鹰界，凌萍萍、汪庆 7；无锡马山，凌萍萍1006；镇江焦山，凌萍萍、汪庆2673.

222.*Taxiphyllum anmoriense* (Resch.) Iwats.中国 -日本成分

句容茅山，李登科20325；宝华山，李登科20240，凌萍萍、汪庆1548-b；苏州虎丘，凌萍萍 67；镇江焦山，凌萍萍、汪庆2664，2679；南山，凌萍萍2755.

223.*Vesicularia ferriei* (Card.et Ther.) Broth.中国 -日本成分

常熟虞山，李登科17503；宜兴小黑沟，吴鹏程、金岳杏2390,凌萍萍、汪庆2121; 吴县东山，吴鹏程、金岳杏1702，凌萍萍699，674；西山，吴鹏程、金岳杏1855；连云港柳河，凌萍萍、汪庆3157；南京中山植物园，凌萍萍941，1852.

224.*Eurohypnum leptothallum* (C.Muell.) Ando 亚洲温带成分

宜兴乾元界，吴鹏程、金岳杏2024；龙池，吴鹏程、金岳杏2333; 溧阳平桥，李登科18934；吴县西山，吴鹏程、金岳杏1763，1789，1808，1817, 1825；东山，吴鹏程、金岳杏1732；无锡蠡园，吴鹏程、金岳杏1971; 句容茅山，凌萍萍、汪庆2343，2375，2349，凌萍萍093.

225.*Ctenidium capillifolium* (Mitt.) Broth.中国 -日本成分

宜兴馨山，吴鹏程、金岳杏2443，邓懋彬7091a; 乾元界，吴鹏程、金岳杏2046；老鹰岩，凌萍萍、汪庆2017；苏州天平山，陈邦杰115；无锡鼋头渚，吴鹏程、金岳杏1932；常熟，李登科17627.

Rhytidiaceae垂枝藓科

226.*Okamurae hakoniensis* (Mitt.) Broth.中国 -日本成分

苏州虎丘，凌萍萍768；常熟虞山，李登科17598; 溧阳平桥,李登科18911.

Diphysciaceae短颈藓科

227.*Diphyscium fulvifolium* Mitt.

宜兴小黑沟, 高彩华26810, 26882; 龙池, 凌萍萍、汪庆2189; 茗岭, 秦怀兰106-3; 吴县西山，吴金等1831.

Polytrichaceae金发藓科

228.*Atrichum undulatum* (Hedw.) P.Beauv. 北温带成分

苏州虎丘，凌萍萍、汪庆2592，2591；灵岩山，凌萍萍、汪庆2638；吴县光福官山，凌萍萍、汪庆2378；句容茅山, 凌萍萍、汪庆2344；无锡惠山公园，凌萍萍汪庆2277；南京中山植物园，凌萍萍1664；连云港连岛，凌萍萍、汪庆3075；柳河，凌萍萍348，354；前云台山，凌萍萍418.

229.*Atrichum crispulum* Schimp. ex Besch.

苏州天平山，凌萍萍、汪庆2609;吴县,光福香雪海，凌萍萍554；无锡市植物园，凌萍萍999；昆山亭林公园，凌萍萍、汪庆3308；南京中山植物园，凌萍萍115；连云港柳河，凌萍萍、汪庆3120；宿城，凌萍萍、汪庆3161；花果山，凌萍萍、汪庆3241，凌萍萍473.

230.*Polytrichum commune* Hedw.世界广布成分

南通军山，凌萍萍、汪庆2910，2912.

231.*Polytrichum formosum* Hedw.北温带成分

吴县东山白河山，凌萍萍648；无锡市植物园，凌萍萍1018；吴县西洞庭湖缥渺山，吴鹏程、金岳杏1750，1796.

232.*Pogonatum inflexum* (Lindb.) Lac.温带亚洲成分

无锡鼋头渚，凌萍萍、汪庆2289，惠山，凌萍萍、汪庆2275；吴县东山，凌萍萍汪庆2512；连云港后云台山，凌萍萍、汪庆2968；大桅尖一宿城途中，凌萍萍、汪庆3106；宜兴龙池，凌萍萍、汪庆2188；老鹰界，凌萍萍、汪庆2003；句容茅山，凌萍萍、汪庆2345.

233.*Pogonatum akitense* Besch.中国 –日本成分

常熟虞山，凌萍萍、汪庆2261；句容茅山，凌萍萍、汪庆2353；吴县东山，凌萍萍、汪庆2509；南京江浦滴水珠，凌萍萍1644；连云港中学，凌萍萍534；宿城，凌萍萍398；无锡马山，凌萍萍962.

234.*Atrichum rhystophyllum* (C.Muell.) Par.

宜兴老鹰界，吴鹏程、金岳杏2186；句容茅山，凌萍萍063.

235.*Pogonatum spinulosum* Mitt.

连云港后云台山，凌萍萍408，436；花果山，凌萍萍270，287；宿城,凌萍萍140.

浙江开化古田山自然保护区苔藓植物区系研究

田春元
(湖北孝感师专生物系，湖北，432100)

A STUDY OF THE BRYOFLORA OF GUTIANSHAN NATURE RESERVE IN KAIHUA COUNTY, ZHEJIANG PROVINCE, EAST CHINA

Tian Chun–yuan
(Biology Department of Xiaogan Teacher's College, Hubei 432100, China)

Abstract: 325 species of bryophytes are reported for Gutianshan Nature Reserve, in Kaihua County, Zhejiang Province.The study area is situated 29°20'N and 118°14'E and is 1246m in elevation. The reported bryophytes include 89 species of liverworts in 39 genera and 22 families and 236 species of mosses in 103 genera and 33 families. Comparisons of the bryoflora of Wanmulin Nature Reserve and those of other nearby mountains such as Huangshan, Tianmu-shan, Sanqing-shan and Wuyanling Nature Reserve show that the floristic relationship between the bryoflora of Gutianshan and Sanqinshan is closest with a similarity coefficient of 66.98% at the generic level.
Key words: Bryophytes; Floristic relationship; Gutianshan Nature Reserve; E China

摘要: 作者于1992年10月和1993年5月两次赴古田山调查，共采集苔藓植物标本1164号．经初步鉴定,共计325种(含种以下单位)．其中苔类39种,隶属于22科和39属；藓类236种，隶属于33科和103属．

古田山苔藓植物区系地理成分复杂，以东亚成分为主，欧亚一北美成分和旧世界热带成分都对其有重要影响．古田山苔藓与日本苔藓植物关系密切．
关键词: 苔藓植物区系；古田山自然保护区；浙江

一、前言

古田山位于中国浙江省西北部，地处中亚热带北部，山势险峻，岩石峋嶙，沟谷纵横，峰峦叠嶂，地形复杂，形成多种多样的小气候环境，受夏季风影响甚大，有明显的季节性变化．植物区系丰富，林分结构复杂，具有中亚热带森林自然特征，因而该地区苔藓植物区系极其丰富多样，加之古田山素有“浙境真胜地”之美誉，故古田山已于1975年3月被浙江省政府正式批准建立为“古田山自然保护区”(省级)，总面积20515亩．

因以前没有人对该地区的植物区系进行系统的调查研究，该方面的资料甚为缺乏，苔藓植物区系的资料尤其如此，根本就没有苔藓植物方面的资料见诸报导，这种情况对于一个自然保护区来说是难以接受的，因而，对该地区苔藓植物区系进行全面的调查研究，不仅将会填补这一重大空白，而且还会为浙江省苔藓植物区系乃至全国苔藓植物区系的研究及《中国苔藓植物志》的编写工作提供基本的素材．鉴于此，作者于1992年10月和1993年 5月两次赴古田山进行苔藓植物的野外调查和采集工作，共采集苔藓植物标本1164号．本文主要就是在整理和鉴定这些标本的基础上，对古田山苔藓植物的区系组成和地理成分等方面进行的分析和研究．

二、自然概况

1.地理位置、地形地貌

古田山位于浙江省西北部的开化县境内，距县城60公里，西面与江西省婺源县、德兴县毗邻，属南岭山系怀玉山脉的一部分．主峰青尖，海拔1246米，北纬29°20'，东径118°14'.

古田山孕育于中生代侏罗纪，距今约有2亿年的历史，由于白垩纪的燕山运动,发生多次的火山爆发而形成现在以花岗岩为主的山体．山脉呈东北一西南走向．花岗岩侵入体岩性坚硬，富垂直节理，经风化形成悬崖峭壁多处，地形复杂，地势险要，山溪源短流急，比降大．山地坡度一般在30°左右，最大约70°．古田山由三条主岗和两条大沟组成，水流经苏庄折入江西境内，经江西德兴县的乐安江流入中国最大的淡水湖鄱阳湖，再汇入长江水系．

2.气候

该区地处中亚热带季风型气候带，受夏季风影响较大，一年中气候有明显的季节性变化，又由于山势高峻，地形复杂，河谷深切，加之整个山体呈东北一西南走向，形成北、东、西三面群峰环抱，对北方冷气团起到阻挡作用，而南方的暖湿气团则由西南的山口遇回入内，因而形成了独特的多种多样的小气候环境．总的特点是冬暖夏凉，温暖湿润，水量充沛，尤其是绝对保护区内茶湾和柏树坑两条大沟，在严重干旱季节也泉水不断．古田山无气象站，但根据距古田山仅50余公里(直线距离)的开化县城气象站1964 -1981年的气象资料统计，年平均温度为16.3°C，最热月平均温度为27.90°C, 最冷月平均温度为4.1°C, 极端最高温度为41.3°C, 极端最低温度为-11.2°C. 日照时数1747.5小时．相对湿度80.9%．年降雨量1778.2毫米，最高达2563.5毫米．年降雨天数140天，主要集中在4 - 6月．无霜期约为250天．

3.土壤及植被

⑴土壤：古田山的土壤类型主要是红壤和黄壤．海拔500 -700米以下为红壤．700-1000米为黄红壤,1000米以上为红黄壤．另外,古田庙一带局部低洼处(海拔850米-1000左右)有山地草甸土分布．土壤呈酸性反应，pH值在5.5-6.5之间，该山到处都有白发藓科的植物和里白科的芒萁分布，正好说明了这点．由于茂密的森林、复杂的植被类型，形成深厚的腐殖质层，肥力较高，有机质含量丰富，表土层呈褐色，土壤富含钾．本地区基本上属于中亚热带山地红黄壤类型．

⑵植被：根据《中国植被》，古田山属于中亚热带北部亚地带湿润常绿阔叶林区域．由于古田山地处偏僻，林茂山险，交通不便，人为活动少，故1958年以前，整个古田山的森林生长十分繁茂，从山麓到山顶一片葱绿，蔚为壮观．但1958年以后，由

于开辟了从星口至苏庄毛坦的公路，建立了伐木场，人为活动显著加强，大力开采森林的结果，海拔450米以下大多沦为荒山灌丛，加之100多年来遭受多次森林火灾，距今最近的一次约在本世纪20年代，因而目前古田山林相保存较为完好的多在海拔400米以上的柏树坑及茶湾的上半部.

总之，古田山优越的地理、气候、土壤和森林植被环境，为苔藓植物生长和分布创造了得天独厚的条件，使得该地区的苔藓植物种类极其丰富多样.

三、小结

㈠ 古田山苔藓植物种类丰富，东亚特有种较多；区系成分以东亚成分为主体，欧亚一北美成分和旧世界热带成分都占有相当的比重；与日本，尤其是本州以南地区的苔藓植物之间的关系极为密切.

㈡ 古田山苔藓植物区系与浙江西天目山和安徽黄山关系密切，与浙江乌岩岭关系介于黄山和三清山之间.

㈢ 对古田山苔藓植物的进一步深入研究，将有助于了解全国苔藓植物不同区系成分间的南迁北移问题.

本文是在导师胡人亮教授的悉心指导下完成的，在标本的整理和鉴定过程中，始终得到王幼芳老师、朱瑞良老师的指导和帮助．上海自然博物馆刘仲苓老师也为本文的撰写提供了很好的资料．在此，一并对他们致以衷心的感谢和诚挚的敬意！

附录：浙江开化古田山苔藓植物名录

Appendix: The catalog of bryophytes of Gutian shan in Kaihua County, Zhejiang Provincc

HEPATICAE

1. Herbertaceae
   1. *Herbertus huerlimannii* H.A.Miller. 18630
   2. *H. aduncus* (Dicks.) Gray 19010, 18640
   3. *H. aduncus* (Dicks.) Gray fo. minor K.C.Chang 18949
   4. *H. fragilis* (Steph.) Herz. 18863
   5. *H. angustissimus* (Herz.) H.A.Miller. 18956,18268
2. Pseudolepicoleaceae
   6. *Blepharostoma minus* Horik. 17892
3. Lepidoziaceae
   7. *Bazzania tridens* (Reinw. & al.) Trev. 17867,18047, 18011,18016, 18302,

18475,18505,18487,18489,18414,18531,17943,18424,18898,18303,18136

8. *B. yoshinagana* (Steph.) Steph. 17891, 18280, 18250, 18270,18355,17968, 17985,18100,18473,18583,18582,18916,18888,18875

9. *B. japonica* (Sande Lac.) Lindb. 17875

4. Calypogeiaceae

10.*Calypogeia muelleriana* (Schiffn.) K.Muell.18266,18271

11.*C. arguta* Nees & Mont. 18741,18648,18448

12.*C. tosana* (Steph.) Steph. 18677,18413,18440

5. Cephaloziaceae

13.*Cephalozia leucantha* Spruce. 18651

14.*C.ambigua* Mass. 17988

15.*C. dtaruensis* Steph. 18699,18532,18537,18545,18546,18749,18712

16.*Nowellia curvifolia* (Dicks.) Mitt. 18663,18664,19671,18674,18658,18656, 18644,18643,18356,18357,18366

17.*Odontoschisma denudatum* (Nees) Dum. 18662,18644,18643

6. Cephaloziellaceae

18.*Cephaloziella microphylla* (Steph.) Douin 18739,18669,18631,18787,18782

19.*C. recurvifolia* Hatt. 17862

7. Jungermanniaceae

20.*Jungermannia atrovirens* Dum. 18409,18528,17906,18365

21.*J. infusca* (Mitt) Steph. 18556,18549,18843,18839,18837

22.*J. breviperianthia* Gao 18010

23.*Mylia verrucosa* Lindb. 18714,18833

24.*Notoscyphus lutescens* (Lehm. & Lindenb.) Mitt. 18779

8. Lophoziaceae

25.*Jamesoniella autumnalis* (DC.) Steph. 18742, 18731, 18659, 18651,18557, 18545,18544,18537,18532

26.*Chandonanthus birmensis* Steph. 18675, 18622

9. Scapaniaceae

27.*Diplophyllum serrulatum* (K.Muell.) Steph. 18857,18844,18872,18835,18836, 18372

28.*Scapania bolanderi* Aust. 19011, 19004

29.*S. stephanii* K.Muell. 18667, 18927, 18396

10.Geocalycaceae

30.*Heteroscyphus argutus* (Reinw. et al.) Schiffn. 18203

31.*H. planus* (Mitt.) Schiffn. 18300, 18428, 18426, 18066, 18373, 18312

11.Plagiochilaceae

32.*Plagiochila arbuscula* (Brid. ex Lehm. & Lindenb.) Lindenb. 17986.

33.*P. sciophila* Nees 17871, 18259, 18402

12.Radulaceae

34.*Radula cavifolia* Hampe 18657,18646,18638,18637,18624

35.*R. japonica* Gott. ex Steph. 18371,18911,18416,18465,18212,18210,18215

36.*R. kojana* Steph. 18012

37.*R. lindenbergiana* Gott. ex Hartm. f.17954

38.*R. obtusiloba* Steph. 18305

39.*R. oyamensis* Steph. 18971, 18297, 18384, 18377,18517,17994,18211,18257

13.Porellaceae

40.*Macvicaria ulophylla* (Steph.) Hatt. 18445,18340,18229,18341

41.*Porella caespitans* (Steph.) Hatt. var. nipponica Hatt. 18464

42.*P.campylophylla* (Lehm.et Lindenb.) Trev. 18331,18333

43.*P.chinensis* (Steph.) Hatt. 18229

44.*P.pinnata* L.18325

45.*P.perrottetiana* (Mont.) Trev. 18293

14.Frullaniaceae

46.*Frullania amplicrania* Steph. 18788

47.*F. aoshimensis* Horik. 18675, 18725

48.*F. apiculata* (Reinw. et al.) Dum. 18140,18523,18963

49.*F. davurica* Hampe 18219,18206,18205

50.*F. densifolia* Steph. ex Evans 18834,18823,18814,18813,18815,18789

51.*F. inflata* Gott. 18539, 18767, 18550

52.*F. inflexa* Mitt. 18762, 18765

53.*F. moniliata* (Reinw. et al.) Mont. 17970

54.*F. muscicola* Steph. 18680, 18852, 18541, 18558,18469,18368,18390,18391, 18392, 18217, 18225, 18230, 18231

55.*F. nepalensis* (Spreng.) Lehm. & Lindenb. 18031

15.Lejeuneaceae

56.*Cheilolejeunea imbricata* (Nees) Hatt. 18515, 18510, 18277

57.*Drepanolejeunea erecta* (Steph.) Mizut. 18750

58.*Lejeunea catanduana* (Steph.) Miller. et al. 18527, 18272, 18415

59.*L. flava* (Sw.) Nees 18851, 18383, 18856, 18427, 18850

60.*L. japonica* Mitt. 18284, 18273

61.*L. parva* (Hatt.) Mizut. 18051

62.*L. ulicina* (Tayl.) Gott. et al.18970, 18083, 18408, 18455,18456,18458, 18636, 18803, 18797, 18850

63.*Cololejeunea appressa* (Evans.) Bened. 18415

64.*C. pseudofloccosa* (Horik.) Bened. 18385

65.*Lopholejeunea applanata* (Reinw. et al.) Schiffn. 17894

66.*L. subfusca* (Nees) Steph. 18975, 18297, 18964

67.*Ptychanthus striatus* (Lehm. et Lindenb.) Nees 18314

68.*Spruceanthus semirepandus* (Nees) Verd. 18345,18352,18376,18386, 18864, 18847, 18855, 18593, 17948, 18308, 17996, 18364, 18294, 18299

69.*Trocholejeunea sandvicensis* (Gott.) Mizut. 18769, 18358, 18337, 18338, 18385,18228,18227,18103, 18223, 18950

16.Makinoaceae

70.*Makinoa crispata* (Steph.) Miyake 18987

17.Aneuraceae

71.*Aneura pinguis* (L.) Dum. 18115, 18173, 17987

72.*Riccardia latifrons* (Lindb.) Lindb. 18411, 18143

73.*R. chamaedryfolia* (With.) Grolle 18965, 18967

74.*R. multifida* (L.) S.Gray. 18969, 18501, 18467, 18195, 18193

18.Metzgeriaceae

75.*Metzgeria furcata* (L.) Dum. 18311, 18389

19.Wiesnerellaceae

76.*Dumortiera hirsuta* (Sw.) Nees 18126

20.Conocephalaceae

77.*Conocephalum japonicum* (Thunb.) Grolle 18670, 18194, 18176

21.Aytoniaceae

78.*Plagiochasma rupestre* (Forst.) Steph. 18724, 18184, 18244

79.*Reboulia hemisphaerica* (L.) Raddi 18498

22.Marchantiaceae

80.*Marchantia polymorpha* L. 18172, 18134

81.*M. nepalensis* Lehm. & Lindenb. 18717, 18733

82.*M. emarginata* Reinw. et al. subsp. *tosana* (Steph.) Bischl. 18960,18481, 18479, 18461, 18183, 18183, 18128, 18142, 18146, 18961, 18591, 18592,18162

83.*M. convoluta* Gao & K.C.Chang. 18504, 18198, 18171, 18163, 18129, 17977

MUSCI

23.Sphagnaceae

84.*Sphagnum palustre* L. 18045

24.Fissidentaceae

85.*Fissidens adelphinus* Besch. 18351, 18379, 18446, 18434, 18679, 18412

86.F. dubius P. Beauv. 18291, 18221, 18292,18201,18060,18525,18589,18476, 18419, 18962, 18992, 18884

87.*F. gymnogynus* Besch. 18144, 17885, 17979

88.*F. geminiflorus* Dozy & Molk. 18101, 18313

89.*F. zippelianus* Dozy & Molk. 17981

90.*F. oblongifolius* Hook. f. & Wils. 18730,18737,18740,18748,18763,18772, 18796, 18790, 18063, 18418, 18447

91.*F. perdecurrens* Besch. 18125

92.*F. grandifrons* Brid. 18503, 18500

25.Ditrichaceae

93.*Ditrichum difficile* (Duby) Fleisch. 17935

94.*D. heteromallum* (Hedw.) Britt. 18443,18437

95.*D. pallidum* (Hedw.) Hampe 18613, 18641,18650,18704,18436,18453,18454, 18421, 18792, 18040, 18350

96.*D. pusillum* (Hedw.) Hampe 17967

97.*Ditrichopsis gymnostoma* Broth. 18781, 18735, 18729,18869,18711,18780, 18832, 17905

98.*Ceratodon purpureus* (Hedw.) Brid. 18989

26.Dicranaceae

99. *Bruchia sinensis* Chen 17860

100.*Campylopus schmidii* (C.Muell.) Jaeg. 18727

101.*C. fragilis ssp. zollingerianus* (C. Muell.) J.-P. Frahm 18703

102.*C. subulatus Schimp.* 18701, 17918

103.*C. japonicus* Broth. 17911, 18335

104.*C. ericoides* (Griff.) Jaeger. 18159

105.*C. hemitrichus* (C.Muell.) Jaeger. 18907, 18242, 18668

106.*C. pyriformis* (Schultz.) Brid. 18616, 18626

107.*C. schimperi* Milde 18625, 18615

108.*C. umbellatus* (Arn.) Par. 18037, 17864, 18247,18926,18153,18002,18518

109.*C. atro-virens* De Not. 18005, 17863, 17854

110.*C. setifolius* Wils. 19009

111.*Dicranodontium attenuatum* (Mitt.) Wils. ex Jaeg. 17917

112.*D. filifolium* Broth. 18778

113.*D. fleischerianum* Schultz-Motel 18914, 17919

114.*Atractylocarpus alpinus* (Milde) Lindb. 18620

115.*Brothera leana* (Sull.) C.Muell. 18655

116.*Amphidium lapponicum* (Hedw.) Schimp. 18269

117.*Symblepharis vaginata* (Hook.) Wijk & Margadant 18889

118.*Holomitrium densifolium* (Wils.) Wijk & Marg. 17916, 17865

119.*H. cylindraceum* (P.Beauv.) Wijk & Marg. 18860,18859,18864,18775

120.*Dicranum hokinense* (Besch.) Gao 18439

121.*D. assamicum* Dix. 18050

122.*D. thelinotum* C.Muell. 18794

123.*D. gymnostomum* Mitt. 18315

124.*D. majus Turner* 18158

125.*D. scoparium Hedw*. 18114, 17874

27.Leucobryaceae

126.*Leucobryum scarbum* Sande Lacoste 18319, 18093, 18966, 18890

127.*L. javense* (Brid.) Mitt. 18008, 18155, 17854, 18486

128.L. bowringii Mitt. 18353, 18354, 18322,18288,18034,18033,17868,17903, 17909, 17940, 18830, 18820, 18706, 18939, 18617, 19003, 18893, 18745,18770

28.Pottiaceae

129.*Gyroweisia brevicaulis* (C.Muell.) Broth. 18065

130.*Oxystegus cuspidatus* (Doz. & Molk.) Chen 18190

131.*O. cylindricus* (Brid.) Hilp. 17930

132.*Tortella tortuosa* (Hedw.) Limpr. 18004, 18161, 18127, 18138

133.*T. humilis* (Hedw.) Jenn. 18090, 18645, 18874,18929,18930,18736,187764

134.*Pseudosymblepharis duriuscula* (Mitt.) Chen 18976

135.*P. angustata* (Mitt.) Chen 18218, 18124, 18483, 18995

136.*Weisia controversa* Hedw. 18560, 18509, 18608, 18698, 18609, 18635

137.*W. crispa* (Hedw.) Mitt. 18404, 17971, 18991

138.*Trichostomum crispulum* Bruch 18988

139.*T. brachydontium* Bruch 18149, 18089

140.*T. involutum* Broth. 18132, 17973

141.*Hyophila involuta* (Hook.) Jaegr. 18178, 18180,18174,18130,18716,18734

142.*H. propagulifera* Broth. 18957

143.*Didymodon fallax* (Hedw.) Zander 18135

144.*D. vinealis* (Brid.) Zander 18519, 18482, 18480, 18472, 18502, 18499,18495

145.*Scopelophila cataractae* (Mitt.) Broth. 18076

146.*Weisiopsis anomala* Broth. & Par. 18375, 18831, 18868, 18747

29.Grimmiaceae

147.*Schistidium apocarpum* (Hedw.) Bruch & Schimp. 18900

148.*S. strictum* (Turn.) Loeske ex O.Maert 18977, 18937,18936,18605,18938, 18246, 18239, 18237

30.Funariaceae

149.*Funaria hygrometrica* Hedw. 19013

31.Bryaceae

150.*Plagiobryum zierii* (Hedw.) Lindb. 18990

151.*Brachymenium nepalense* Hook. 17923, 18468, 18690

152.*B. ptychothecium* (Besch.) Ochi 18405, 18280,18154,18160,18600,18695,18899

153.*Pohlia acuminata* Hoppe & Hornsch. 18179, 18185

154.*P. flexuosa* Hook. 18934

155.*P. longicollis* (Hedw.) Lindb. 18003

156.*Bryum argenteum* Hedw. 18274, 18410, 18607, 18606, 18958, 18980, 18985, 18719

157.*B. calophyllum* R. Brown 18191, 18169, 18166

158.*B. caespiticium* Hedw. 18360, 18403, 18334, 18224, 18168, 18167, 18165

159.*B. capillare* Hedw. 18170, 18507, 18493, 18492,18508,18951,18526,18516

160.*B. lonchocaulon* C. Muell. 18361, 18979

161.*B. cyclophyllum* (Schwaegr.) B.S.G.18118

162.*B. neodamense* (Itzig.) in C. Muell. 18131

163.*B. setschuanicum* Broth. 18119, 18182

164.*B. filiforme* var. *concinnatum* (Spruc.) Boulay 18580

32.Mniaceae

165.*Trachycystis microphylla* (Dozy & Molk.) Lindb. 18996,18997,18999,18014, 18222, 18263, 18335, 18320

166.*M. lycopodioides* Schwaegr. 18919

167.*M. laevinerve* Card. 18697, 18694, 17978

168.*Plagiomnium cuspidatum* (Hedw.) T.kop. 18981,18692,18715,18086, 18085, 18232

169.*P. acutum* (Lindb.) T.Kop. 17980, 18601, 18255

170.*P. rostratum* (Schrad.) T.Kop. 18925, 18902, 18881, 18877, 19007, 19000, 18691, 18521, 18518, 17982, 18141, 18069, 18116, 18310, 18281, 18282, 18256, 18289, 18380

33. Rhizogoniaceae

171.*Pyrrhobryum dozyanum* (Lac.) Manuel 19014

34.Bartramiaceae

172.*Bartramia halleriana* Hedw. 18113, 18318

173.*B. pomiformis* var. *elongata* Turner 18584, 18588

174.*Philonotis bartramioides* (Griff.) Griffin 18238

175.*P. capilliformis* J.-S. Lou & Wu 18187

176.*P. fontana* (Hedw.) Brid. 18496, 18079, 18189, 18241

177.*P. setschuanica* (C. Muell.) Par. 17974

178.*P. thwaitesii* Mitt. 18511

179.*P. turneriana* (Schwaegr.) Mitt. 17999, 17897,18192,18328,18243,18207,18251

35.Ptychomitriaceae

180.*Ptychomitrium dentatum* (Mitt.) Jaegr. 18878, 18906

181.*P. polyphylloides* (C.Muell.) Par. 18940, 18120, 18562, 18942

182.*P. linearifolium* Reim. & Sak. 17872, 17913, 18432, 18946, 18974

183.*P. wilsonii* Sull. & Lesq. 18275, 18240, 18431, 18943, 18728

184.*Glyphomitrium acuminatum* Broth. 18430, 18708

36.Orthotrichaceae

185.*Macrocoma tenue* ssp. *sullivantii* (C.Muell.) Vitt 17941
186.*Macromitrium cylindrothecium* Nog. 18470
187.*M. sinense* Bartram 18758, 18858, 18793, 17910
188.*M. japonicum* Doz. & Molk. 18666, 18721, 18621, 18710, 17922, 18080
189.*M. gymnostomum* Sull. & Lesq. 18766, 18756, 18743, 17946, 17900,17882, 18075, 18027, 18406
190.*M. ferriei* Card. & Ther. 18682, 18777, 18774,18935,18941,18688,18451, 18561, 18529, 17915, 17907, 17851, 18043, 18058
191.*Schlotheimia grevilleana* Mitt. 18348
192.*S. pungens* Bartram 18862, 18994, 18948, 17904, 18029, 18030, 18317

37.Racopilaceae

193.*Racopilum aristatum* Mitt. 18398, 18234, 18084, 17956

38.Hedwigiaceae

194.*Hedwigia ciliata* (Hedw.) Ehrh. ex P.Beauv.18527, 18425, 18720, 18722

39.Cryphaeaceae

195.*Sphaerotheciella sphaerocarpa* (Hook.) Fleisch. 17920
196.*Pilotrichopsis dentata* (Mitt.) Besch. 17898, 17992, 18639,18628,18623,18611
197.*P. robusta* Chen 18876

40.Leucodontaceae

198.*Leucodon secundus* (Harv.) Mitt. 18683

41.Trachypodaceae

199.*Trachypus bicolor* Reinw. & Hornsch. 18059, 18848, 17955, 18110
200.*T. humilis* Lindb. 18892, 18909, 18082, 17939,17961,17964,17893,17878, 18054, 18073, 18071, 18067, 18025, 18254, 18332, 18236
201.*Pseudospiridentopsis horrida* (Card.) Fleisch. 18028

42.Meteoriaceae

202.*Papillaria feae* C.Muell. ex Fleisch. 18944, 18344, 18343
203.*Meteorium taiwanense* Nog. 18908
204.*M. latiphyllum* J.-S.Lou 18952
205.*M. subpolytrichum* (Besch.) Broth. 18905, 18245, 18214, 18326
206.*M. subpolytrichum* ssp. *horikawae* (Nog.) Nog. 17899, 18891
207.*Meteoriopsis squrrosa* (Hook.) Fleisch. 18062
208.*Chrysocladium retrorsum* (Mitt.) Fleisch. 19005, 19001, 17953, 17957, 17959, 18145, 18024, 18370, 18604, 18038, 18661, 18685
209.*Floribundaria floribunda* (Doz.& Molk.) Fleisch. 18462
210.*F. sparsa* (Mitt.) Broth. 17853, 18046
211.*Aerobryidium filamentosum* (Hook.) Fleisch. 17886, 17857
212.*A. wallichii* (Brid.) Fleisch. 17976, 17852, 18262
213.*Aerobryum speciosum* Dozy & Molk. 18306, 18285, 18362

214.*Aerobryopsis parisii* (Card.) Broth. 18490
215.*A. subdivergens* (Broth.) Broth. 17975, 17887, 18243, 18321
216.*Barbella flagellifera* (Card.) Nog. 18433, 18579, 18587, 18565, 18307
217.*B. pendula* (Sull.) Fleisch. 18922, 18921, 18164, 18123
218.*B. trichophora* (Mont.) Fleisch. ex Broth. 18323
219.*B. niitakayamensis* var. propagulifera (Nog.) S.-H.Lin 18112
220.*Pseudobarbella angustifolia* Nog. 18098
221.*P. validiramosa* Wu & J.-S.Lou 18594
222.*Meteoriella soluta* (Mitt.) Okam. 18634
43.Neckeraceae
223.*Homaliodendron flabellatum* (Sm.) Fleisch. 17880
224.*H. scalpellifolium* (Mitt.) Fleisch. 18923,18912, 18474, 17888, 17858, 17856,17855,18096,18309,18216,18260,18286
225.*Homaliadelphus targionianus* (Mitt.) Dix.& P. de la Varde 18913,18845, 18810,17942,17944,17950,18061,18057,18301,18283,18388
226.*Thamnobryum subseriatum* (Mitt. ex Lac.) Tan 18290,18401
44.Hookeriaceae
227.*Hypopterygium japonicum* Mitt. 17889,18478,18477,19006
45.Leskeaceae
228.*Pseudoleskeopsis zippelii* (Dozy & Molk.) Broth. 18393,18955
229.*Bryonoguchia molkenboeri* (Lac.) Iwats. & Inoue 18181
46.Thuidiaceae
230.*Haplohymenium flagelliforme* Savicz-Ljubitskaya 18258
231.*H. sieboldii* (Doz.& Molk.) Dozy & Molk. 17966
232.*H triste* (Ces.) Kindb. 18018,17949,17921,18799, 18042, 18053, 18070, 18894,18865,18854,18853,18811, 18374,18367, 18072,18812,18808, 18785, 18696,18689,18681,18761,18947
233.*Claopodium aciculum* (Broth.) Broth. 18278,18064, 18538, 18540, 18553, 18513,18954
234.*C. assurgens* (Sull. & Lesq.) Card. 18684
235.*C. prionophyllum* (C. Muell.) Broth. 18597,18596,18598,18599
236.*Haplocladium angustifolium* (Hampe & C.Muell.) Broth. 18339
237.*H. microphyllum* (Hedw.) Broth. 18151,18148,18570,18578, 18612, 18983, 18982,18986,18718,18723,18686
238.*Thuidium philibertii* Limpr. 18381
239.*T. cymbifolium* (Doz. & Molk.) Doz. & Molk. 18121
240.*T. glaucinum* (Mitt.) Bosch & Lac. 18316,18220,18590,18497,18660,18882
241.*T. glaucinoides* Broth. 18276, 18196, 17879, 17890, 17989,17983,17951, 18543,18554,18555,18972,18973

47.Anomodontaceae

242.*Anomodon giraldii* C.Muell. 18783,18776,18760,18757,18755,18752, 18744, 18709

48.Amblystegiaceae

243.*Amblystegium serpens* (Hedw.) B.S.G. 17963,17962

244.*A. tenax* (Hedw.) Jens. 18235

49.Brachytheciaceae

245.*Palamocladium nilgheriense* (Mont.) C. Muell. 19002,18524,18233,18485, 17945,18296,17958,17952,18013

246.*Brachythecium auriculatum* Jaegr. 17960, 18494

247.*B. plumosum* (Hedw.) B.S.G. 18678

248.*B. helminthocladum* Broth. & Par. 18879

249.*B. velutinum* (Hedw.) B.S.G. 18095

250.*B. reflexum* (Stark.) Schimp. 18520

251.*B. campestre* (C.Muell.) B.S.G. 19012

252.*B. rutabulum* (Hedw.) B.S.G. 18654,18574,18687

253.*B. pendulum* Tak. 18571,18568

254.*B. brotheri* Par. 18466, 18920

255.*Rhynchostegium pallidifolium* (Mitt.) Jaegr. 18394,18471,18806

256.*Rhynchostegiella sakuraii* Tak. 17931

257.*R. curvisteta* (Brid.) Limpr. 17877

50.Entodontaceae

258.*Entodon macropodus* (Hedw.) C. Muell. 18945,18484,18463, 18226, 18330, 18097,18068

259.*E. viridulus* Card. 18295

260.*Mesonodon flavescens* (Hook.) Buck 18984

51.Plagiotheciaceae

261.*Plagiothecium denticulatum* (Hedw.) B.S.G. 17965

262.*P. laetum* B.S.G. 18265

263.*P. neckeroideum* var. *sikkimense* Ren. & Card. 18552

264.*Pseudotaxiphyllum pohliaecarpum* (Sull. & Lesq.) Iwats. 18359, 18204, 18044, 17895,17927, 17928,18816, 18809,18444,18441,18422,18959,18873, 18932,18838,18842,18726

52.Sematophyllaceae

265.*Struckia argentata* (Mitt.) C.Muell. 18642

266.*Wijkia tanytricha* (Mont.) Crum 17901

267.*Chionostomum rostratum* (Griff.) C.Muell. 18530,17914

268.*Pylaisiadelpha erythrocaulis* (Mitt.) Buck 18395,18248,18009,18870

269.*P. integrifolia* (Broth.) Buck 18378,18346,18347, 18304, 18298, 18202,

18188,17883,18795,18647,18652,18968

270.*P. falcatula* (Broth.) Buck 18147,18705,18896,18676

271.*Warburgiella cupressinoides* C.Muell. ex Broth. 18673,18784,18786

272.*Sematophyllum henryi* (Par. & Broth.) Broth. 17938

273.*S. pulchellum* (Card.) Broth. 18840,18109,18015

274.*S. caespitosum* (Hedw.) Mitt. 18800,18801,18049,18108,18022

275.*S. robustulum* (Card.) Broth. 18208,18209

276.*S. phoeniceum* (C.Muell.) Fleisch. 18407,18732,18417,18457,18512,18438, 18422,18449,18450,18819,18408,18572,18563,17947,18150

277.*S. subhumile* (C.Muell.) Fleisch. 18771,18768,18759,18738,18753,18713, 18791, 18567,17991,17993,17995,18048,18102,18020,18342

278.*S. subhumile* ssp. japonicum (Broth.) Seki 18821, 18822, 18825, 18802, 18807,18586,17937,17990

279.*S. humile* (Mitt.) Broth. 17902,17936,18026

280.*S. micans* (Mitt.) Braithw 18817,18818,18824,18841

281.*Acroporium turgidum* Mitt. 18122

282.*Rhaphidostichum boscii* ssp.thelidictyon (Dozy) Seki 18998

283.*Taxithelium kerianum* (Broth.) Broth. 18514

284.*T. nepalense* (Schwaegr.) Broth. 18137

53.Hypnaceae

285.*Giraldiella levieri* C.Muell. 18055, 18036

286.*Pylaisiella brotheri* (Besch.) Iwats. 18039

287.*Homomallium incurvatum* (Brid.) Loeske 18019

288.*Hypnum plumaeforme* W.Wilson 18566,18569,18573,18577,18585,18506,18702, 18917,18904,18897,18993,18880,18886,18665

289.*H. cupressiforme* Hedw. 18363,18397,18399,18264,18261,18117,18197,18186, 18035,18092,18106,18056,17884,17908,17929,18547,18805,18629,18267

290.*H. pallescens* (Hedw.) P. Beauv. 18152,18094,18139,17870,17912,18933

291.*H. oldhamii* (Mitt.) Jaegr. 17876,18000

292.*H. hamulosum* Schimp. 18252,18032,18001

293.*H. tristo-viride* (Broth.) Par. 17869

294.*H. revolutum* (Mitt.) Lindb. 18074

295.*H. callichroum* Brid. 18105

296.*H. sakuraii* (Sak.) Ando 18400,18088,18746

297.*H. subimponens* Lesq. 18866

298.*H. sikkimense* Ando 18861

299.*Pseudostereodon procerrimum* (Mol.) Fleisch. 18707

300.*Ectropothecium buitenzorgii* (Bel.) Mitt. 18077

301.*E. ohosimense* Card. & Ther. 18700

302.*Trachythecium verrucosum* (Jaeg.) Fleisch. 18023

303.*Isopterygium albescens* (Hook.) Jaegr. 18533,18534,18535,18536, 18827, 18826,18804,18798,18871,18910

304.*I. minutirameum* (C.Muell.) Jaegr. 18382,18751

305.*Taxiphyllum taxirameum* (Mitt.) Fleisch. 18199,18177,18087,18133,18052, 17969,18460

306.*T. maniae* (Ren. & Card.) Fleisch. 17984

307.*Vesicularia reticulata* (Doz. & Molk.) Broth. 18915

308.*Ctenidium molluscum* (Hedw.) Mitt. 18576

309.*C. capillifolium* (Mitt.) Broth. 18369,18327,18491,18885,18903

310.*C. elegantulum* Broth. 18924

54.Rhytidiaceae

311.*Rhytidiadelphus triquetrus* (Hedw.) Warnst. 18522

55.Polytrichaceae

312.*Atrichum crispulum* Schimp. ex Besch. 18557

313.*A. undulatum* (Hedw.) Beauv. 18928,18091,18693,18602,18603,18429,18595, 18551,18548,17934,17998,17866,18078,18006,18175

314.*A. undulatum* var. *gracilisetum* Besch. 18632,18633,18542

315.*Pogonatum aloides* (Hedw.) P.Beauv. 18619,18610,18867,18672,18828,18107

316.*P. neesii* (C.Muell.) Dozy 18627,18099,18157,18041

317.*P. inflexum* (Lindb.) Lac. 18978,18593,18614,18423,18081,18887, 18895, 18931, 18349, 17873, 17896,17972,17933,17932,17929,17924,18007,18329, 18324

318.*P. cirratum* ssp. *fuscatum* (Mitt.) Hyvonen 18918, 18488, 17881, 18111, 18249,18287

贵州梵净山藓类植物新纪录*

林齐维
(贵州省环境保护科学研究所，贵阳 550002)
吴鹏程
(中国科学院植物研究所，北京 100093)

MOSSES NEW TO MT. FANJING, GUIZHOU PROVINCE, SW CHINA

Lin Qi–wei
(*Institute of Environmental Science of Guizhou Province, Guiyang, Guizhou 550002, China*)
Wu Pan–cheng
(*Institute of Botany, Academia Sinica, Beijing 100093, China*)

Abstract: The southwest province of Guizhou is located on the second-step of Chinese geo-topography. Its western and eastern borders connect the first and third geotopographical steps.Mt.Fanjing occupies a strategic position in the northeastern corner of the province. The mountain has a well marked vertical vegetational zonation and is rich in bryophyte species.Since 1960,B.-G.Zhong and S.-Z.Jian has reported more than 760 species of mosses from the mountain. In 1993, the authors participated in a Sino-American expedition to Mt. Fanjing and discovered 11 genera and 16 species representing new records of bryophytes for the province. In this paper,another 2 genera and 5 species of bryophytes are reported new to Mt. Fanjing.
Key words: Mt. Fanjing; Guizhou; Bryophytes; Sino-American Expedition

摘要：贵州地处中国二级台地，西侧及东侧分别与三级和一级台地交界，苔藓植物种类极其丰富．位于贵州东北部的梵净山，系武陵山脉的主峰，垂直差异明显，为苔藓植物提供极其良好的生育环境．80年代以来，鍾本固、姜守忠等多次报道该区苔藓植物，其中藓类植物总计 186种．1991年，笔者组织中美苔藓植物考察，再度在梵净山进行采集．经初步研究鉴定，为梵净山增加藓类植物11属，16种新纪录．
关键词：梵净山；藓类植物；新纪录．

梵净山为武陵山脉主峰，横亘于贵州东北部松桃、印江、江口三县交界处，为黔东北第一高山．地理位置为北纬27°46'50"–28°1'30"，东经108°35'55"–108°48'30"．梵净山自然保护区东西宽21km，南北长27km，总面积38743ha．最高峰凤凰山海拔2570.5m，全区相对高差达2000m以上，垂直分布明显．在梵净山的西南和西北坡，

* 国家自然科学基金资助项目
This project is supported by the National Natural Science Foundation of China.

海拔800m–1400m为北亚热带；1400–2000m为南温带(暖温带)，而2000m以上为中温带．该山东北坡和东南坡的海拔600m以下为中亚热带，600–1250m为北亚热带，1250–1900m为南温带，在1900m以上则为中温带．梵净山区总的气候特征为中亚热带季风山地湿润气候．梵净山在地质构造上属武陵古陆梵净隆起，形成于前震旦系，经中生代燕山运动、新生代喜马拉雅运动，尤其是第四纪新构造运动，而使该山区成为高于周围的穹窿断块山地．以最高峰为中心，有马槽河、黑湾河、盘溪沟、凯土河、牛尾河、肖家河、淘金河、余家沟等八条河流呈放射状向四周流出，侵蚀剧烈，切割深，比降大，山峦起伏，坡陡谷深，地形复杂．垂直高度的显著差异和复杂的地形，使梵净山区植物种类丰富，植被复杂．

本世纪30年代，蒋英(1930)、焦启元(1931)、锺补勤(1937、1938)等，在梵净山作种子植物调查时，附带采集了苔藓植物标本，是梵净山苔藓植物调查采集的最早纪录．焦启元所采标本，经Edwin B. Bartram研究鉴定，于1935年在《中国藓类植物区系补遗》(Additions to the moss Flora of China)中报道藓类植物41种、

1963－1964年,简焯坡等在调查梵净山水青冈林时，也附带采集苔藓植物标本．1979年10月及1980年11月，贵阳师范学院生物系姜守忠带领学生教学实习，专门进行了苔藓植物的调查采集，主要地点为梵净山东南坡，包括江口县黑摩河－铜矿厂－鱼坳、茴香坪、金顶、老金顶－芙蓉坝－印江县金盏坪－盘溪河．1981年5月,屠玉麟等在考察梵净山植被时采集了苔藓植物标本．姜守忠、屠玉麟等所采标本，经锺本固、姜守忠研究鉴定，报道了梵净山苔藓植物50科，115属和245种．其中，苔类20科，27属和59种；藓类30科，88属和186种(1983，1990)．

梵净山藓类植物的新记录

白发藓科Leucobryaceae

* *Leucobryum bowringii* Mitt. 23553

真藓科Bryaceae

** *Anomobryum filiforme* (Dicks.) Solms var.*concinnatum* (Spruce) Loeske 23721

丛藓科Pottiaceae

* *Didymodon rigidicaulis* (C.Muell.) Saito 23902

桧藓科Rhizogoniaceae

** *Rhizogonium dozyanum* Lac. 23461

扭叶藓科Trachypodaceae

* *Duthiella flaccida* (Card.) Broth. 23654

蔓藓科Meteoriaceae

* *Floribundaria aurea* (Mitt.) Broth. 23480

平藓科Neckeraceae

* *Homaliadelphus targionianus* (Mitt.) Dix. et P. de la Vard. 23643

* *Homaliodendron exiguum* (Bosch et Lac.) Fleisch. 23585

* *Neckera yezoana* Besch. 23443

** *Isothecium subdiversiforme* Broth. 23442

** *Thamnobryum sandei* (Besch.) Iwats. 23468

薄罗藓科Leskeaceae

** *Pseudoleskeopsis japonica* (Sull.) Iwats. 23577

碎米藓科Fabroniaceae

* *Fauriella tenuis* (Mitt.) Card. 23389

羽藓科Thuidiaceae

* *Haplocladium microphyllum* (Hedw.) R.Watanabe et Iwats. 23490

* *Thuidium kanedae* Sak. 23419

短颈藓科Diphysciaceae

** *Diphyscium fulvifolium* Mitt. 23541

** 属的新记录

* 见B.C.Tan等1994年的报道

参考文献

[1] 钟本固、熊源新,1989, 贵州藓类植物名录(I),贵州师范大学,自然科学版10: 41–51.

[2] 钟本固、熊源新,1990,贵州藓类植物名录(Ⅱ), 贵州师范大学,自然科学版14:22–32.

[3] 钟本固、熊源新, 1990b, 梵净山苔藓植物初步研究, 周政贤主编: 梵净山研究 275–305页.

[4] Bartram, E.B., 1935, Additions to the moss flora of China. *Ann. Bryol.* Ⅷ: 6–21.

[5] Brotherus, V. F.,1929, In H.Handel-Mazzetti (ed.) Symbolae Sinicae Ⅳ:Musci, pp.1–147. Wien.

[6] Tan, B.C, Q. W. Lin, M.R. Crosby et P.C.Wu, 1994, A Report on the 1991 Sino-American Bryological Expedition to Guizhou Province, China: New and Noteworthy Addition of Chinese Moss Taxa. *The Bryologist* 97(2):127–137.

江西五倍子蚜虫冬寄主(藓类植物)研究 *

季梦成
(江西农业大学园艺系，南昌 330045)

A STUDY OF THE WINTER HOSTS (MOSSES) OF GALLNUT APHIDS IN JIANGXI PROVINCE, E CHINA

Ji Mengcheng
(Jiangxi Agriculture University, Nanchang, 330045,China)

Abstract: Gallnut is a special production in Southeast Asia. It has a wide range of uses and is a famous medicinal material in China.This paper reports on the winter hosts (18 species of mosses in 10 genera) of 3 species gallnut aphids in Jiangxi Province.The reported number of winter hosts (mosses) constitute 47.4% of the total number of moss species serving as winter host plants for gallnut aphids. Detailed distribution information of these mosses is provided to serve as a scientific basis for the increase production of gallnut aphids.

Key words: Gallnut aphids; Mosses; Winterhosts; Jiangxi Province, E China

摘要：五倍子为东南亚特产，具有广泛的应用领域，中国尤以药用著名．本文报道了江西产3种倍蚜虫，即角倍蚜、蛋铁倍蚜和倍花蚜．倍蚜冬寄主植物（藓类）计有10属和18种，占全国已知该类植物的47.4%，文中详细记述了其种类及分布，为江西五倍子生产提供依据．

关键词：五倍子；藓类植物；江西

五倍子为瘿绵蚜科(Pernphigidae)的某些蚜虫寄生在漆树科(Anacardiaceae)盐肤木属(*Rhus*)一些植物上所形成的虫瘿，是蚜虫一树木一藓类生物圈中的产物，为东南亚特产，但朝鲜、日本等地产量甚微．北美产一种五倍子,其倍蚜虫为美国倍蚜(*Melaphisrhois* Fitch)，它的生活史与中国倍蚜虫几乎相似．中国的五倍子产量居世界首位，年产量4000吨左右，且品质最优，历来是一项重要的出口商品．中国对五倍子的文献记载，应算晋人郭璞(公元276-342年)的《山海经》为最早，距今有1700多年的历史．

五倍子单宁在金属防腐、稀有金属的提取，合成纤维印染固色、食品工业、医药卫生、三废处理和化学分析等方面已得到广泛的应用．中国五倍子自古即以药用著称,主要用作收敛剂，有消肿、消毒、防腐、化淤、止血、镇痛等功效．

---

* 江西省自然科学基金资助项目．
This project is supported by the JiangXi Natural Science Foundation of China.
本文承中国科学院昆明植物所黎兴江研究员审阅，并得到江西省药物所张海道研究员、江西农业大学俞志雄教授的帮助，谨致衷心感谢!

中国五倍子的主产省份为贵州、四川、湖南、湖北、陕西及云南等6省,产量约占全国总产量的90%以上，上述6省中，常年收购量100吨以上的地区(州)有11个,25吨以上的县有44个．江西亦为五倍子生产省份，年均收购量仅19.5吨，而贵州为1329.8吨，四川848.7吨，湖北694.2吨，湖南375.3吨，陕西210.8吨,云南119.2吨．生产五倍子必须同时具备五倍子致瘿蚜、夏寄主及冬寄主三个条件，需有一个虫一树一藓的良性循环，江西大部分地区具备五倍子(特别是角倍)适生气候条件，但长期以来， 由于历史，产供销经营体制等诸多原因，五倍子生产一直处于“自然生长、个人采收、零星收购”的状况．

一、江西五倍子蚜虫、夏寄主的种类及分布

江西共有五倍子蚜虫三种，即角倍蚜(*Schlechtendalia chinensis* Bell.)、蛋铁倍蚜(*Kaburagia ovogallis* Tsai et. Tang)、倍花蚜(*Nurudea shirakii* Mats.)，占全国的21.4%(表1)，江西倍蚜虫种类不多，但分布较广，特别是角倍蚜，全省广布．倍蚜虫的夏寄主为盐肤木 (*Rhus chinensis* Mill.) 及红麸杨 [*R. punjabensis* var. *sinica* (Dicls.) Rchd.]二种，占全国的2/3，其中盐肤木全省广布，红麸杨分布于江西东北部的德兴、景德镇、婺源、铅山等地．

表1.江西五倍子蚜虫的种类、分布及夏寄主

倍蚜虫种类	分布	夏寄主植物
角倍蚜	全省广布	盐肤木
蛋铁倍蚜	赣西北、赣南	红麸杨
倍花蚜	赣东北、赣南、永修、武宁、黎川	盐肤木

二、江西五倍子蚜虫冬寄主(藓类植物)的种类及分布(表2.)

表2.江西五倍子蚜虫冬寄主植物的种类及分布

冬寄主植物	倍　蚜	冬寄主植物分布	备注
疣灯藓 (*Trachycystis microphylla*)	角倍蚜	庐山、永修、武宁、宜丰、赣南、德兴、铅山、黎川	
长尖提灯藓 (*Mnium lycopodioides*)	角倍蚜	赣西北、赣南、宜春、婺源、景德镇、贵溪	
缘边匐灯藓 (*Plagiomnium acutum*)	角倍蚜	全省广布	

树形匐灯藓 (*P.arbuscula*)	角倍蚜	永修、遂川、崇义、大余、龙南、赣东北	*(1)
日本匐灯藓 (*P.japonicum*)	角倍蚜	赣东北、吉安、井冈山、上犹、崇义	*(2)
侧枝匐灯藓 (*P.maximoviczii*)	角倍蚜	全省广布	
具喙匐灯藓 (*P.rhynchophorum*)	角倍蚜	全省广布	
大叶匐灯藓 (*P.succulentum*)	角倍蚜	全省广布	
灰白青藓 (*Brachythecium albicans*)	蛋铁倍蚜	全省广布	
多褶青藓 (*B.buchananii*)	蛋铁倍蚜	赣东北、赣西北、宁岗、遂川、井冈山	
卵叶青藓 (*B.rutabulum*)	蛋铁倍蚜	赣东北、永修、新建	
密叶尖喙藓 (*Oxyrrhynchium óavatieri*)	蛋铁倍蚜	全省广布	
东亚金灰藓 (*Pylaisia brotheri*)	倍花蚜	永修云居山	
大灰藓 (*Hypnum plumaeforme*)	倍花蚜	全省广布	
尖叶灰藓 (*H. callichroum*)	倍花蚜	全省广布	
鳞叶藓 (*Taxıphyllum taxirameum*)	倍花蚜	全省广布	
狭叶绢藓 (*Entodon angustifolius*)	倍花蚜	赣东北、武宁、修水、永修、宜丰靖安	
美灰藓圆枝变种 (*Eurohypnum leptothallum* var. *tereticaule*)	倍花蚜	九江庐山	

*(1) 黎兴江(1986)认为该种不能作为角倍蚜的冬寄主，而张海道(1992)认为此种为江西角倍蚜适宜冬寄主，因作者在野外调查时，未见该种上的越冬倍蚜，故尚待进一步调查、观察.

*(2) 张海道(1992)提出：“江西的角倍蚜基本上不在上面越冬”．符合唐觉(1976)认为此种不能作为角倍蚜的冬寄主，而张满祥(1984)则记录本种为五倍子蚜虫冬寄主,此亦尚待进一步调查再作定论.

三、讨 论

1.五倍子生产对环境条件要求较严格，要求虫(致瘿蚜)树(夏寄主)藓(冬寄主)三者配套生长，缺一不可．江西部分地区，如鄱阳湖盆地、吉泰盆地，植被条件较差，常常有林无藓或有藓无林，且夏季高温持续时间长，不宜作为五倍子发展区，江西大部分山区和丘陵地区，生态环境较好，常有成片分布的盐肤木及大量的提灯藓，宜作为五倍子发展区．

2.江西产五倍子最主要是角倍，全省广布．其夏寄主盐肤木，冬寄主多种提灯藓亦全省广布，三者容易配套，有着广阔的生产前景．因此，在我省广大地区，进行人工抚育营造盐肤木林，并因地制宜在各地筛选出1–2种相配套的提灯藓，作为角倍蚜的最佳冬寄主，对其进行人工繁殖，可达到生产并大幅度提高江西五倍子产量的目的．

3.到目前为止，已知江西五倍子蚜虫有3种，种类占全国的21.4%，冬寄主藓类植物有10属18种，种类占全国的47.4%,随着五倍子资源调查，及苔藓植物研究工作的进一步开展，可以肯定，江西五倍子蚜虫及冬寄主藓类植物的种类会比目前有所增加．积极开展倍蚜虫及冬寄主种类的调查，适生倍蚜的引种、驯化及放养试验，开展冬寄主藓类植物的人工大面积培殖研究, 促使生产与科研相结合, 将丰富我省五倍子品种，促进五倍子产量和品质的提高．

参考文献

[1] 陈邦杰等，1963、1978: 中国藓类植物属志(上、下册)，科学出版社，北京．

[2] 黎兴江(主编)，1985: 西藏苔藓植物志. 科学出版社，北京．

[3] 黎兴江、臧穆，1979: 中国提灯藓科的研究. 云南植物研究，1(1):32–80．

[4] 吴鹏程等，1984: 苔藓名词及名称. 科学出版社，北京．

[5] 向和，1985: 中国五倍子的地理分布和主要倍蚜虫生态条件初步分析.动物世界，12(1):45–53．

[6] 贵州林科所编，1986: 中国五倍子研究报告选编．

[7] 季梦成，1993: 赣北云居山苔藓植物的初步研究. 江西农业大学学报，15(2):174–181．

[8] 李勇民，1994: 北美的五倍子. 贵州农学院学报，13(1):24-27．

[9] Hu Renliang,1983:A Revision of the Chinese species of Entodon.*The Bryologist*. 86(3):193–233.

[10] Koponen, T,1981: A Synopsis of Mniaceae (Bryophyta). Ⅵ. Southeast Asian taxa. *Acta Bot. Fennica* 117:1–34.

广西苔藓植物药用资源(一)

杜泽乡
(桂林医学院，桂林 541001)

A STUDY OF MEDICINAL BRYOPHYTES USED IN GUANGXI PROVINCE, S CHINA

Du Ze-xiang
(Guilin Medical College, Guilin, 541001, China)

Abstract: Medicinal bryophyte resource of Guangxi province is reported based on field investigation. 20 species were reportedly used as medicine. The curative effect, distribution and ecology of each of the medicinal bryophytes used locally are provided.
Key words: Guangxi Province; Medicinal bryophyte resource

摘要: 本文首次报道了中国广西产苔藓药用植物资源20种．并阐述了它们的地理分布、生境及主要功效．
关键词: 广西; 苔藓药用植物; 资源

广西位于北纬20°54'-26°24'，东经104°24'-112°04'，总面积23.75Km^2，属热带向亚热带过渡带地区，气候适宜，雨量丰富，蕴育了大量野生药用植物，现知约有600种，但广西产苔藓药用植物，迄今未见报道．笔者通过多年野外调查和查阅整理大量苔藓植物标本及资料，就广西产苔藓药用植物资源的种类和地理分布，现报道20种．

一、广西苔藓药用植物

波叶仙鹤藓*Atrichum undulatum* (Hedw.) P. Beauv. 全区各地有分布，生于山地阴湿林边、路旁土坡．能抗癌．

直叶珠藓*Bartramia ithyphylla* Brid. 百色、合池有分布，生于海拔1500-2000m的林下阴湿处．能镇惊安神；治心慌心烦、癫痫、中风不语．

银叶真藓*Bryum argenteum* Hedw. 全区广布，生于土墙、房顶、火烧林地．能清热解毒；治细菌性痢疾、鼻窦炎．

万年藓*Climacium dendroides* (Hedw.) Web.et Mohr.分布桂西南，生于低洼积水林地、溪涧．能清热除湿、舒筋活络．

蛇苔*Conocephalum conicum* (L.) Dumort. 全区广布,生于林下湿地、沟谷岩石上．能清热解毒、消肿止痛；治疗痈肿、蛇咬伤、无名肿毒、烫伤、刀伤、骨折．

牛角藓*Cratoneuron filicinum* (L.) Loesk. 分布桂西南，生于林下石上、树干上．

黄牛毛藓*Ditrichum pallidum* (Hedw.) Hamp. 分布百色，生于土坡、空旷林地．能

镇惊、治小儿惊风.

扁枝绢藓*Entodon compressus* (Hedw.) C. Muell. 桂南有分布，生于潮湿岩石、树干、土上．能利尿、治水肿．

葫芦藓*Funaria hygrometrica* Hedw.分布全区各地，生于田圃、居住处周围、林地肥沃湿土．能消炎止痛、除湿止血、祛风通络；治鼻窦炎、跌打损伤、肺热吐血．

尖叶小羽藓*Haplocladium microphyllum* (Hedw.)Broth.全区大部分地区有分布，生林地阴湿处，或附生树兜、腐木上．能消炎止痛、退热；治咽喉炎、气管炎、肺炎、胃肠炎．

尖叶提灯藓*Mnium cuspidatum* Hedw.全区广布，生于溪边、阴湿土坡、树兜．能凉血止血；治鼻衄、崩漏．

地钱*Marchantia polymorpha* L. 广布全区，生于阴湿土坡、石壁上．能清热拔毒，消炎生肌；治刀伤、骨折、毒蛇咬伤、疖痈肿毒．

东亚金发藓*Pogonatum inflexum* (Lindb.)Par.分布于桂北，生林边或路旁．能镇静安神；治神经衰弱．

大金发藓*Polytrichum commune* Hedw. 全区大部分地区有分布，生山野阴湿处．能滋阴补虚、清热解毒、凉血止血；治肺病咳嗽、吐血、刀伤出血、盗汗．

石地钱*Reboulia hemisphaerica* (L.) Raddi全区广布，生岩石及土坡上．能消肿止血；治外伤出血、跌打肿痛．

暖地大叶藓*Rhodobryum giganteum* (Hook.) Par.桂北有分布，生溪边、潮湿林地．能清心明，发神；治冠心病．

泥炭藓*Sphagnum cymbifolium* Ehrh.及暖地泥炭藓*S.junghuhnianum*分布区内各地，生沼泽地．能清热，代药棉用．

小石藓*Weisia viridula* Hedw.三江等地有分布，生空旷地，岩面．能清热解毒、治鼻炎．

鳞叶藓*Taxiphyllum taxirameum* (Mitt.) Fleisch.区内各地有分布,生林下湿地．能止血消炎;治外伤出血．

二、结论

广西苔藓植物种类数以百计，药用种类近50种．全草药用均可，药效十分广泛，从神经系统疾病到心血管疾病，从内科到外科，从抗癌抗菌到消炎止痛，且大部分种类全区有分布，是值得开发的药用植物资源．苔藓植物除药用外，还有许多用途，如保持水土，测定大气污染，指示森林类型，作包装运输新鲜苗木的保湿、果树嫁接包扎接口，播种后的覆盖物，作肥料及燃料等，是值得研究和利用的资源．但由于苔藓植物个体小，体积轻，作为药物，需要量较多，可考虑引种栽培以满足需求．另外，苔藓药用植物的有效成分、药理、临床效用等均值得深入研究．

参考文献

[1] 李树刚等编，1988，《广西植物资源》：80页．北京科技出版社．

一个奇异的化石叶—北票叶（新属）

段淑英 王鑫
（中国科学院植物研究所）

A PECULIAR FOSSIL LEAF-*BEIPIAOPHYLLUM* DUAN & WANG, GEN NOV.*

Duan Shuying and Wang Xin
(*Institute of Botany, Academia Sinica, Beijing, 100093,China*)

Abstract: A peculiar fossil leaf with *Bennettiales*-type leaflets and with a slender petiole is described. The leaflet is similar to that of the genus *Nilssoniopteris* in external form and cuticle structure, but the palmately compound leaf is special. Since these specimens could not be classified in any known genus, a new genus, *Beipiaophyllum*, is proposed.
Key Words: Bennettitales; *Beipiaophyllum*; Late Triassic; Liaoning province, NE China

摘要: 本文描述了一个特殊的化石叶．化石采自中国东北辽宁省北票市附近羊草沟村的晚三叠世地层中．该地层和其中的化石已有许多人做过研究．此化石也被定为不同名称：周惠琴（1981）定为拟丹尼蕨（未定种）*Danaeopsis sp.*; 米家榕等人(1993)定名为羊草沟带羊齿（*Taeniopteris yangcaogouensis*）．我们仔细研究后，不仅找到一块掌状复叶标本，而且在极困难的情况下研制出一小块表皮．根据化石叶片的外部形态和表皮细胞特征，我们认为该化石既不是拟丹尼蕨，也不属于带羊齿，它的表皮细胞特征与蕉带羽叶(*Nilssoniopteris*)最接近，所以其分类位置应属于本内苏铁纲，因与蕉带羽叶不同而定为一新属：北票叶属（*Beipiaophyllum*）属型种为羊草沟北票叶*B. yangcaogouensis*（Mi et al.）Duan & Wang comb. nov.
关键词: 本内苏铁; 北票叶; 晚三叠世; 辽宁省

Introduction

We collected many fossil leaves from a village named Yangcaoguo,near the coal city Beipiao,western part of Liaoning province, NE China. The fossil assemblage consists of about 30 species belonging to *Equisetales*, *Filicopsida*, *Ginkgopsida*,and *Coniferopsida*.The dominant genera among the fossils are *Neocalamites*, *Cladophlebis*, *Pterophyllum*, *Nilssonia*, *Baiera*, *Ginkgoites*,*Podozamites* etc.. The most important species are *Neocalamites carcinoides Harris, N.rugosus Sze, Cladophlebis kaoiana Sze, C.gracilis*

* The project is supported by the National Natural Science Foundation of China and special support of Palaeontological & Palaeoantthropological Investigation from Chinese Academy of Sciences.

Sze, Glossophyllum shensiense Sze, Podozamites lanceolatus (L. et H.) Braun, P.*schenki* Heer,and *Cycadocarpidium erdmanni* Nathorst. The geological age of the fossil-bearing strata was assigned to the Late Trassic, which was not only studied by us but also by palaeobotanists Zhou (1981),Mi and his students (1993),as well as by other geologists.

In the collections we found many single leaves of *Taeniopteris*-type which was designated *Danaeopsis* sp. by Zhou (1981) and later described as a new species: *Taeniopteris yangcaogouensis* Mi et al.(1993). After our carefully study, a strange fossil was discovered. As far as we know no known fossil genus is similar with this fossil leaf in its morphological form and cuticle structure, therefore a new genus was set up.

Description

Genus *Beipiaophyllum* Duan & Wang gen.nov.

Type species:*Beipiaophyllum yangcaogouensis* (Mi et al.) Duan & Wang comb. nov

Diagnosis: Same as for type species.

Derivatio nominis: both the names of genus and species are derived from the city and the village where the fossil was found.

Beipiaophyllum yangcaogouensis (Mi et al.) comb. nov.

1981. *Danaeopsis* sp. Zhou, p. 147, pl. I, fig. 8.

1993. *Taeniopteris yangcaogouensis* Mi et al., p.157, pl.51, figs. 7-10, text-fig. 39.

(P1. I, figs.1-3, P1. II, figs. 1-2, text-figs. 1, 2)

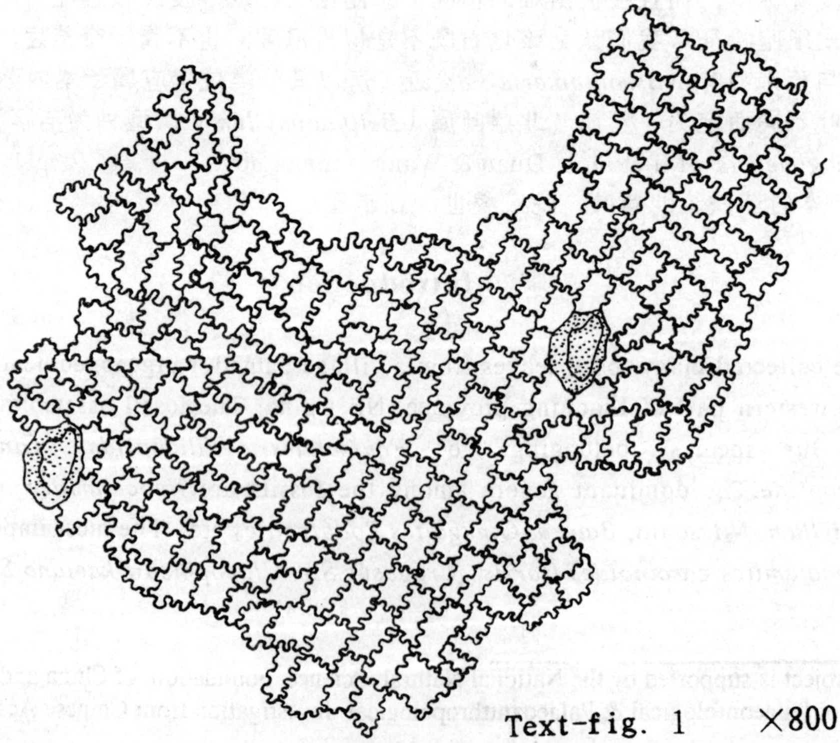

Text-fig. 1 ×800

Emended diagnosis. Palmately compound leaf or clustering leaves consisting of 8-9 leaflets. Each single leaflet is *Taeniopteris*-type, with a slender petiole 1.5-2mm wide and 20-40mm long.Margin of leaflet entire or slight undulate. Midrib conspicuous from the base to the apex, lateral veins simple or forked 1-2 times near the midrib or margin The vein concentration 7-8 per cm near the midrib, 10-12 per cm near the margin. Cuticle structure *Nilssoniopteris*-type, with strong sinuous cell walls.

Holotype: No.8905.

Paratype: No.8907. 8919.

Description: Single leaflet *Nilssoniopteris*-type, 50-120 mm long and 20-50 mm wide, widest part near the apex, contracting rapidly towards the apex and slowly towards the base. Apex obtuse or subacute.Margin entire or slight undulate. Venation *Nilssoniopteris*-type, midrib conspicuous, 1.5-3 mm to 1mm from leaf base to the apex, the lateral veins simple or forked once or twice near the midrib or margin, arising at a concentration of 7-8 per cm near the midrib, 10-12 per cm near the margin. A specimen (pl.II, fig.2) shows three leaflets with slender petioles growing from one point, another specimen (pl.II, fig. 1) with slender petioles converging toward one point. *Taeniopteris yangcaogouensis* Mi et al.showed a similar specimen(pl.51, fig.10, Mi et al., 1993).Only one big specimen shows a complete palmately compound leaf consisting of 8-9 leaflets (pl.I,fig.1), which is assigned as the holotype of this species. Substance of lamina rather thin, the cuticle very delicate and hard to prepare. We got a very small piece of cuticle from the big specimen (pl.I, fig.1), showing very marked cell character (pl.I, fig. 2-3, text-fig.1), and strong sinuous cell walls. Some rectangular cells forming about four lines may follow the vein; other epidermal cells polygonal, rectangular or square. Among them are two trichome bases. No stomata on this piece of cuticle.

Comparison and discussion. Most specimens of this species were preserved as a single leaflet, *Taeniopteris*-type in morphological form and venation characters. Some specimens show slender petiole. The substance of lamina is rather thin, so Mi et al. assigned it as a new species of *Taeniopteris*. But the clustering or palmately compound leaf is distinct from any species of *Taeniopteris*.The similar fossil *Cladotaeniopteris shaanxiensis* Zhang & Mo.from the Lower Permian of Shaanxi, North China showed a clustering leaves at the tip of very short branch and axis covered with irregularly small round scars. The leaflet form is taeniopterid, with very short petiole (5 mm long and 5-3 mm wide), but the venation is obscure,only showing midrib,the secondary veins either indistinct, the cuticular structure unknow.So this fossil is different from ours. It was suggested to have cycadean affinity,also different from our species. *Nilssoniocladus nipponensis* (Yokoyama) Kimura (1975) from the Early Cretaceous of Japan,showed long and short shoots and a terminal cluster leaves at the tip of short shoots, but the leaf form

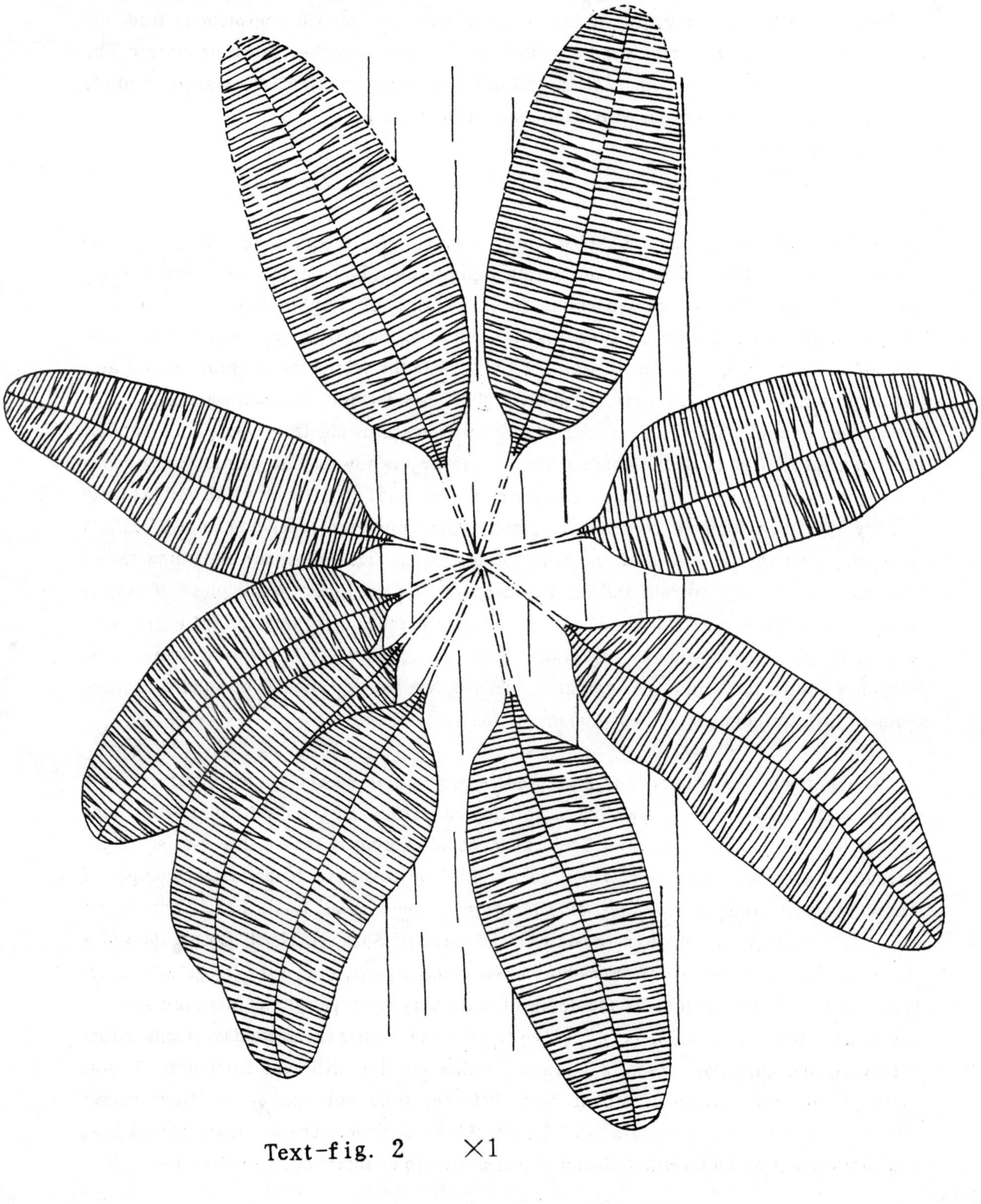

Text-fig. 2 ×1

and venation are *Nilssonia*-type, very different from our specimens. There is no cuticle character from the Japanese species. So our specimens could not be put into either of the two genera. Our specimen is similar to the genus *Nilssoniopteris* on the leaflet's external form, venation and cuticle structure, even we have only a minuter piece of cuticle which appears to be a upper cuticle, because upper cuticle normally is thicker than the lower one, and there is no stomata on the upper cuticle. But in the genus of *Nilssoniopteris*, there is no clustered leaves or palmately compound leaf. (text - fig.2). Therefore we would rather set up a new genus than placing it as a new species of *Nilssoniopteris*.

According to the morphology and cuticular structure, this species cannot be assigned in *Taeniopteris* nor in *Nilssoniopteris*.The taxonomic position of the new genus must be close to *Nilssoniopteris* belonging to Bennettitales.

Acknowledgements

The authors are deeply indebted to the senior geological enginner Jiankang Lu and his young colleague Yunfeng Bi of the Beipiao coal mine,who helped us collect the fossils. We are also grateful to Mr.Z. Du for the photes, and Miss X.Xu for drawing the text-figures.

References

[1] Harris,T.M.,1969, The Yorkshire Jurassic flora.III,Bennettitales. British Museum (Natrual History) London, PP.61–79.

[2] Kimura, T., and Sekido, S., 1975, *Nilssoniopteris* n.gen.(Nilssoniaceac n. Fam.) newly found from the Early Lower Cretaceous of Japan.Palaeontographica. Abteilung B, Bd. 153. 1-3. p. 111–118.

[3] Mi Jia–rong, Zhang Chuan–bo, Sun Chun–lin et al.1993, Late Triassic stratigraphy, paleontology and paleogeography of the Northern part of the circum-pacific belt, China. Science press, Beijing. P.157–158.

[4] Zhang Shan–zhen & Mo Zhang–guan,1981, On the occurrence of cycadophytes with slender growth habit in the Permian of China. Geological Society of American Special paper 187.

[5] Zhou Hui–qin, 1981, Discovery of the Late Triassic fossil plants from Yangcaoguo, Beipiao, Western Liaoning province. Selected works of the 12th Annual conference of the palaeontological society of China.Science Press.

PIATE I

1

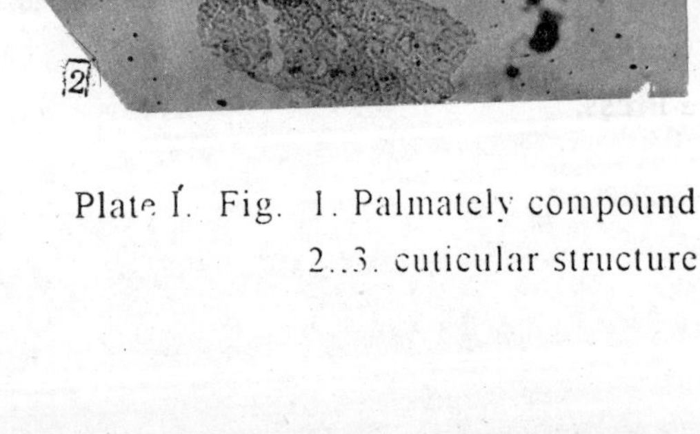

2

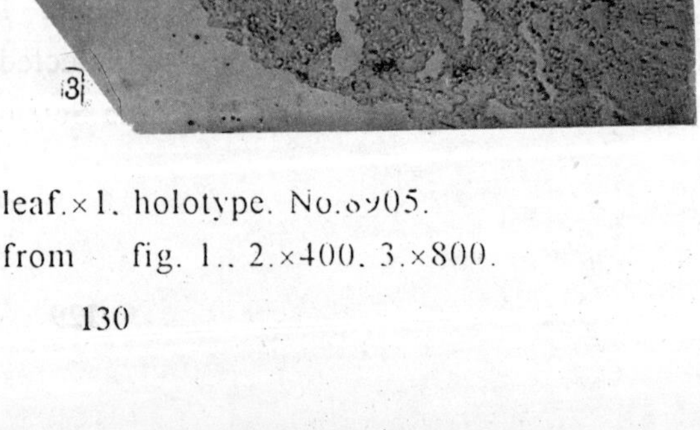

3

Plate I. Fig. 1. Palmately compound leaf.×1. holotype. No.8905.
2..3. cuticular structure from fig. 1.. 2.×400. 3.×800.

PLATE II

2

Plate II

Figs.1.2. Leaflets and it's long slender petiole. ×1. paratype.No. 8907, 8919.

国产耳蕨属一新种

向莉琳

(北京自然博物馆，北京 100050)

A NEW SPECIES OF THE FERN GENUS, *POLYSTICHUM* ROTH (DRYOPTERIDACEAE) FROM CHINA

Xiang Lilin

(*Beijing Natural History Museum, Beijing 100050, China*)

Abstract: A new species of *Polystichum, P. bigemmatum* R. -C. Ching ex L. -L.Xiang, was described from Sichuan Province based on a collection of S.-G. Wu(#2075).

Key words: New species; Fern; *Polystichum; P.bigemmatum*

关键词：新种；耳蕨；双胞耳蕨.

双胞耳蕨 新种 (图版：1–4)

Polystichum bigemmatum Ching ex L.L. Xiang sp. nov.

Species insignis, rhachidis apice bigemmis a speciebus ejusdem sectionis distincta. Planta habito *P.neolobato* proxima, differt fronde apice vivipara, anguste oblonga, 5.1–7.2cm lata, pinnata; pinnis pinnatipartitis subtus paleis peariibus confertim obtectis.

一个突出的种. 因其叶轴顶端生有两个芽胞而与其它已知的同组相近种不同，其植株形态与*P.neolobatum* Nakai相似，但叶顶端胎生，叶片长圆形，宽仅5.1–7.2cm，一回羽状，羽片深裂而可区别.

植株高约68cm，根状茎直立. 叶簇生，叶柄长约13cm,从基部向上密被褐色卵状披针形及披针形鳞片，向叶轴渐稀疏；叶轴顶端着生两个芽胞；叶片长50–53cm，宽5.1–7.2cm，狭长圆形，渐尖头，基部截形，二回羽裂；羽片30–33对，长2.5–3.6cm，宽1.5–2.3cm，狭长圆形或长圆形，顶端刺状，基部上侧平截,下侧阔楔形一回羽裂；裂片长0.3–0.5cm，宽0.1–0.2cm，顶端具刺，边缘近全缘或呈浅波状. 孢子囊群近羽轴着生；囊群盖棕褐色，圆盾形.

Sichuan（四川）：Miannin（冕宁），Yeqin（冶勤），Sanchahe（三岔河），S. K.Wu.（武素功）2075（(Typus, PE），on rocks in dense forest, alt 2600m.

本文承蒙郭木森先生绘图，在此致谢.

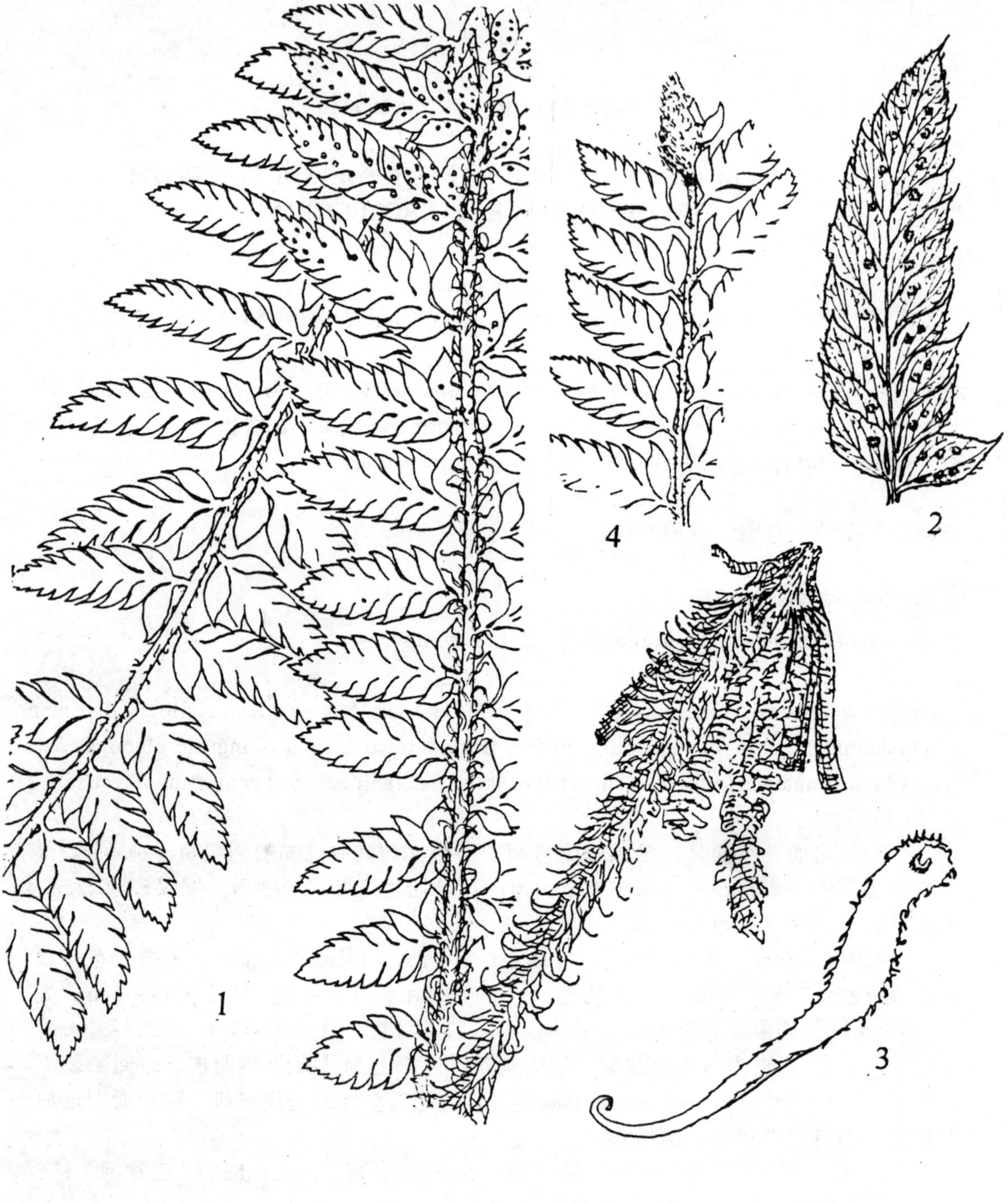

图 1–4 双胞耳蕨 *Polystichum bigemmatum* Ching ex L. L. Xiang

1. 植株的一部分； 2. 羽片及其裂片； 3. 叶柄鳞片； 4.叶轴顶端芽胞(郭木森绘)